RAND NATIONAL DEFENSE RESEARCH INSTITUTE

An Attack Against Them All?

Drivers of Decisions to Contribute to NATO Collective Defense

Anika Binnendijk, Miranda Priebe

Prepared for the Office of the Secretary of Defense
Approved for public release; distribution unlimited

Contents

Tables

Summary

This report provides an analytical framework for understanding allies' willingness to contribute to a military response to Russian attacks on a North Atlantic Treaty Organization (NATO) member. The framework is based on insights from the political science literatures on deterrence, alliance politics, and foreign policy decisionmaking. Specifically, we identify 13 factors that could influence allies' decisionmaking. These fall into three broad categories:

1. **Domestic politics:** Public opposition could restrain national leaders from participating in an operation to counter Russian aggression, but other domestic factors, such as elite consensus and an electorally secure governing coalition, might overcome public opposition.
2. **Perceptions of Russia:** Countries that perceive Russia as ambitious, opportunistic, or threatening to their homelands are more likely to participate, while members with favorable views of Russia or lower threat perceptions are more likely to worry about Russian economic or military retaliation.
3. **Alliance politics:** Countries that place the greatest value on NATO's continuity, for security or nonsecurity reasons, are more likely to participate. However, allies are less likely to join if other major allies—particularly the United States—do not participate, if they perceive allies' goals as divergent, or if they are not confident in their abilities to restrain other allies from unnecessarily escalating a conflict with Russia (see Table S.1 for more details on these categories).

We also consider how the relative importance of these factors might vary in the event of an unconventional Russian attack on a NATO country. We find that the domestic factors outlined above are more likely to dominate in an unconventional scenario. Russia's use of proxies, cyberattacks, and information campaigns could all lead to ambiguity about Russian intentions and about the applicability of Article 5 of the treaty that established NATO. As a result, concerns about the threat that Russia poses or alliance dynamics might be weaker in this scenario. Conversely, in a conventional scenario, when Russian forces are unambiguously engaged in an attack on a NATO ally and Article 5 clearly applies, alliance politics and perceptions of Russian threat considerations are more likely to drive allies' decisionmaking.

Table S.1
Framework for Assessing an Ally's Willingness to Contribute to a Military Response

Categories	Factors That Influence Allies' Decisionmaking
Domestic politics	• Public opinion about military responses and alliance commitments (size of effect is highly context dependent)
	• Vulnerability of governing coalition on foreign policy issues
	• Foreign policy decisionmaking structure
Perceptions of Russia	• Perception of Russian aims and motivations
	• Competing national security demands
	• Vulnerability to Russian military retaliation
	• Vulnerability to Russian economic retaliation
	• Perception of escalation risk
Alliance politics	• Participation of other allies
	• Alignment of goals among participants
	• Ability to restrain coalition members
	• Consequences of alliance dissolution or abandonment
	• Punishment by noncoalition allies

Prior to and during a crisis, Russia would likely take steps to influence the domestic political calculations, alliance considerations, and threat perceptions of individual allies. Drawing from our framework and from previous examples of Russian influence tactics, we identify areas where allies' decisionmaking might be vulnerable to Russian manipulation.

Finally, we propose steps to mitigate Russian influence attempts and increase NATO unity in the event of a Russian attack. For example, we suggest that the United States

- improve U.S. intelligence-sharing on Russian activities by increasing the resources dedicated to collecting and analyzing Russian behavior and releasing intelligence to NATO governments
- enhance public access to information on Russia through financial support to such entities as Radio Free Europe/Radio Liberty, as well as to new NATO institutions, such as the Joint Intelligence and Security Divisions hybrid analysis branch
- initiate new outreach to emerging political elites on NATO's value—e.g., through such institutions as the NATO Parliamentary Assembly
- engage regularly with allies on Russian aims and motivations, including through an update to NATO's Strategic Concept

- address allies' concerns about *entrapment* (the possibility of being dragged into a conflict unnecessarily) by supporting renewed NATO engagement with Moscow and continuing to encourage Baltic states to address minority grievances
- supplement the defensive capabilities of vulnerable allies through U.S. support for additional point defenses and new civil defense exercises and preparations
- mitigate allies' vulnerability to economic coercion by continuing to support diversification of European energy supplies and by initiating a new NATO–European Union strategic reserves program
- establish new ministerial-level political exercises to provoke discussions about NATO goals in a Russia scenario and build allies' confidence in collective decisionmaking.

Abbreviations

AfD	Alternative für Deutschland (Germany)
AWACS	Airborne Early Warning and Control System
EU	European Union
ISAF	International Security Assistance Force
JISD	Joint Intelligence and Security Division
LNG	liquefied natural gas
NAC	North Atlantic Council
NATO	North Atlantic Treaty Organization
SPD	Social Democratic Party (Germany)

CHAPTER ONE

Introduction

The 2018 National Defense Strategy calls for strengthening U.S. alliances and reducing allied vulnerability to coercion in an era of increased great power competition.[1] The North Atlantic Treaty Organization (NATO) was originally founded in 1949 as an institution of collective defense against the Soviet Union. After the collapse of the Soviet Union, NATO's focus shifted to crises near the borders of the alliance (Bosnia and Herzegovina, Kosovo) or further afield to "out of area" missions (Afghanistan, Libya). However, Russia's military incursions into Ukraine in early 2014, coupled with its military buildup and other aggressive behavior, have prompted renewed discussion about the possibility of an attack on a NATO member. As a result of these concerns, NATO has taken practical steps to strengthen deterrence.[2] In spite of these steps, analysts and commentators have raised several questions about allied willingness to respond to an attack on a NATO member, especially against the Baltic states.[3] Some point to polls that show relatively low public support in Europe for defending Baltic allies.[4] Commentators, including former Supreme Allied Commander Europe General

[1] Jim Mattis, *Summary of the 2018 National Defense Strategy of the United States of America: Sharpening the American Military's Competitive Edge*, Washington D.C.: U.S. Department of Defense, 2018.

[2] For example, the 2014 Wales Summit's Readiness Action Plan initiated a new Very High Readiness Joint Task Force, a multinational brigade of 5,000 troops intended to deploy within seven days (NATO, *NATO Wales Summit Guide*, September 3, 2014a). At the 2016 Warsaw Summit, allies established four multinational battlegroups in Estonia, Latvia, Lithuania, and Poland (NATO, "Warsaw Summit Communique," transcript, July 9, 2016). The 2018 Summit launched a NATO Readiness Initiative in which, within the overall pool of forces, allies are to offer an additional 30 heavy or medium maneuver battalions, 30 kinetic air squadrons, and 30 major naval combatants with enabling forces at 30 days' readiness or less by 2020 (NATO, "Brussels Summit Declaration," transcript, July 11, 2018c).

[3] See Christiane Hoffmann and René Pfister, "Part of the West? German Leftists Have Still Not Understood Putin—Interview with Historian Henrich Winkler About Germany and the West," *Spiegel Online*, June 27, 2014; Margriet Drent, Peter van Ham, and Kees Homan, *Article 5 Revisited—Is NATO Up to It?* The Hague, Netherlands: Clingendael, August 2014; Jeffrey A. Stacey, "A Russian Attack on Montenegro Could Mean the End of NATO," *Foreign Policy*, July 27, 2018; Judy Dempsey, "NATO's European Allies Won't Fight for Article 5," *Judy Dempsey's Strategic Europe*, June 15, 2015.

[4] Craig Winneker, "For Europe's NATO Allies, Attack on One Isn't Attack on All," *Politico*, June 10, 2015. For polling data, see Bruce Stokes, "NATO's Image Improves on Both Sides of Atlantic: European Faith in American

Wesley Clark, have suggested that this reflects a widespread cultural aversion to war in Europe.[5] Second, in spite of the emphasis on allies in U.S. strategic documents and from U.S. officials, President Donald Trump has called NATO "obsolete" and raised questions about the U.S. commitment to defending NATO members.[6] Finally, the Russian government is actively sowing divisions among NATO members to prevent unified responses to Russian actions.[7]

In spite of these widespread concerns, there has been little systematic analysis of the factors that would drive allies' decisions to respond to a Russian attack on a NATO member.[8] As a starting point, this report draws from the political science literatures on deterrence, alliance politics, and foreign policy decisionmaking to explore four questions:

- Which factors would drive an ally's decision to contribute to a military response through NATO or as part of a coalition of the willing?
- How might an ally's decisionmaking differ if Russia carried out an unconventional, instead of a conventional, attack?
- What levers does Russia have to influence allies' decisionmaking?
- What steps can NATO members take to promote a unified response and limit vulnerability to Russian manipulation?

The remainder of this introduction will review NATO members' commitment to mutual defense, discuss the types of military responses that NATO members might consider in a Russia scenario, and explain the study's methodology in greater detail.

Military Support Largely Unchanged," Pew Research Center, May 23, 2017.

5 Wesley Clark, Jüri Luik, Egon Ramms, and Richard Shirreff, *Closing NATO's Baltic Gap*, Tallinn, Estonia: International Centre for Defence and Security, May 2016. On changes in European thinking on the use of force, see James J. Sheehan, *Where Have All the Soldiers Gone? The Transformation of Modern Europe*, Boston, Mass.: Houghton Mifflin, 2008.

6 See Michael R. Gordon and Niraj Chokshi, "Trump Criticizes NATO and Hopes for 'Good Deals' With Russia," *New York Times*, January 15, 2017.

7 James B. Foley, "Don't Let Putin Destroy NATO," *Time*, March 31, 2016.

8 Other work has looked at factors affecting ally contributions to past U.S.-led operations but has not addressed the particular challenge of a direct Russian attack on a NATO member (Jason W. Davidson, *America's Allies and War: Kosovo, Afghanistan, and Iraq*, New York: Palgrave Macmillan, 2011; Andrew Bennett, Joseph Lepgold, and Danny Unger, "Burden-Sharing in the Persian Gulf War," *International Organization*, Vol. 48, No. 1, Winter 1994). This report complements a 2018 RAND research report that proposes an exploratory model of factors that might affect national "will to fight" (Michael J. McNerney, Ben Connable, S. Rebecca Zimmerman, Natasha Lander, Marek N. Posard, Jasen J. Castillo, Dan Madden, Ilana Blum, Aaron Frank, Benjamin J. Fernandes, In Hyo Seol, Christopher Paul, and Andrew Parasiliti, *National Will to Fight: Why Some States Keep Fighting and Others Don't*, Santa Monica, Calif.: RAND Corporation, RR-2477-A, 2018). Additional work by the project has applied the model to a NATO conventional conflict within a series of tabletop exercises, with a report forthcoming.

NATO's Commitment to Collective Defense

NATO members' commitment to mutual defense has been the cornerstone of the alliance since it was established by the Washington Treaty in 1949. Article 5 of the treaty commits each individual member state to (1) view an armed attack against one or more members as "an attack against them all" and (2) assist the party under attack by taking "such action as it deems necessary, including the use of armed force" to restore and maintain regional security.[9] Article 5 was originally motivated by concerns that the Soviet Union would seek to expand its control of the European continent. However, there was never a conventional Soviet attack on a NATO member.

The language of NATO's charter preserves a degree of flexibility for allies, stipulating that "each . . . will assist . . . by taking . . . such action as it deems necessary."[10] Members are not required to respond to an attack with armed force. However, for many years, it was a common assumption within the alliance that a conventional attack on any NATO member would elicit a military response from the entire alliance.[11] Although Article 5 can be invoked by NATO government representatives at the North Atlantic Council (NAC) to initiate collective action as an alliance, it can also be invoked by any NATO member, even in the absence of NAC consensus.[12]

To date, the only instance in which NATO has invoked Article 5 was in response to the September 11, 2001, attacks against the United States.[13] Following the invocation, NATO Airborne Early Warning and Control System (AWACS) aircraft flew more than 360 missions over U.S. skies, with more than 830 crew members from 13 allied nations.[14] For allies, including Germany, NATO's Article 5 decision drew alliance considerations to the forefront of public national security discourse.[15] This shift might have contributed to the fact that 15 of 18 NATO allies ultimately contributed forces to NATO operations in Afghanistan, even though it was not an Article 5 mission.[16]

Alliance members can also invoke NATO's Article 4, under which NATO allies will "consult together whenever, in the opinion of any of them, the territorial integrity, political independence or security of any of the Parties is threatened."[17] While

---

[9] NATO, North Atlantic Treaty, Washington, D.C., April 4, 1949.

[10] NATO, "Collective Defence—Article 5," webpage, June 12, 2018b.

[11] Robert E. Hunter, "NATO's Article 5: The Conditions for a Military and Political Coalition," *European Affairs*, Vol. 2, No. 4, Fall 2001.

[12] NATO, 1949.

[13] "Invocation of Article 5 Confirmed," *NATO Update*, October 3, 2001.

[14] Eric Schmitt, "NATO Planes to End Patrol of U.S. Skies," *New York Times*, May 2, 2002.

[15] Christian Tuschhoff, "Why NATO Is Still Relevant," *International Politics*, Vol. 40, No. 1, March 2003.

[16] Tuschhoff, 2003.

[17] NATO, 1949.

largely viewed as a political mechanism, Article 4 consultations have been seen as a step toward invoking collective defense commitments under Article 5 and have sometimes resulted in limited military activity. When Turkey invoked Article 4 during the 2003 U.S. invasion of Iraq, NATO flew AWACS missions and deployed Patriot batteries on Turkish territory for several weeks.[18] NATO allies—including the United States, Germany, Spain and the Netherlands—also deployed Patriot batteries to Turkish soil following Turkey's invocation of Article 4 over the escalating war in Syria.[19] Similarly, the 2014 Russian attack on Ukraine prompted Poland and Lithuania to invoke Article 4.[20] Immediate NATO military responses included a temporary expansion of NATO's Baltic Air Policing mission and the deployment of NATO's Standing NATO Maritime Group 1 to the Baltic Sea.[21] Since that time, allies have bolstered NATO's military presence in Europe's east through the establishment of rotating multinational battlegroups in Poland and each of the Baltic states.[22]

Military Responses

In a hypothetical crisis with Russia, allies would begin consultations under Article 4 or could move immediately to discussions about invoking Article 5. In either case, members would have to consider their political, economic, and military responses. Politically, NATO members could support, abstain from a vote on, or oppose an official NATO response. Should NATO be unable to reach political consensus, each state would need to consider whether to support a coalition of the willing operating outside of the NATO structure.[23] Allies might also consider making financial contributions to support a response.[24]

Each state would also need to decide whether and how to contribute to any military response. A military response does not necessarily involve the use of force; it might involve such action as sending troops to a NATO country to deter an attack. In either

[18] Robert Coalson, "What Are NATO's Articles 4 and 5?" *Radio Free Europe/Radio Liberty*, June 26, 2012.

[19] Michael Moran, "Turkey's Article 5 Argument Finds No Takers," webpage, Carnegie Corporation of New York, February 24, 2016.

[20] "Lithuanian and Polish Presidents Call for NATO Treaty Article 4 Consultations," *EuroDialogueXXI*, March 3, 2014.

[21] Louisa Brooke-Holland, *NATO's Military Response to Russia: November 2016 Update*, London: House of Commons Library, Briefing Paper No. 07276, November 3, 2016.

[22] NATO, "NATO's Enhanced Forward Presence," fact sheet, February 2018a.

[23] Yoel Sano, "Guest Post: Will Russia Make a Play for Estonia, Latvia, and Lithuania?" *Financial Times*, March 23, 2015; Thomas Theiner, "Gotland—The Danzig of Our Time," *Euromaiden Press*, March 22, 2015.

[24] Japan's contribution to coalition operations in the 1990–1991 Persian Gulf War was almost entirely financial (Bennett, Lepgold, and Unger, 1994). Within the NATO alliance, Luxembourg often contributes funding to operations because it has too few military units to commit its troops.

case, an ally might choose not to participate, make a small contribution as a symbolic show of unity, or be a leading member of the military response. However, the magnitude of the contribution is not the only way that allies can tailor their contributions. As discussed in Chapter Two, allies are likely to weigh a number of factors, such as the risk of Russian retaliation, as they decide how to respond. In response to this complex and potentially countervailing set of considerations, allies might try to vary other characteristics of their military contribution, such as whether the contribution is observable, its military utility, and the ease of reversing the military contribution.[25]

For example, a state that that feels strong alliance pressure to contribute but worries about Russian retaliation might be more likely to provide intelligence analysis; such analysis might not be easily detected by Russia, reducing the likelihood of retaliation. Table 1.1 offers illustrative examples of how the nature of a given allied military contribution might vary along characteristics such as observability, cost, military utility, and reversibility.

Research Approach and Report Organization

The core of this study is a framework, presented in Chapter Two, that identifies the range of factors that drive state decisions about whether to contribute to collective defense. To generate this framework, we draw from the deterrence, alliance politics, and foreign policy decisionmaking literatures. In doing so, we drew from both purely theoretical arguments and those that have been tested empirically. In some cases, research spoke directly to the question of allies' decisionmaking about military contributions to collective defense. More often, we extended and applied the logic of existing theories and arguments to the specific problem of a state deciding whether to defend

Table 1.1
Characteristics of Notional Allied Military Contributions

	Observability	Cost	Military Utility	Reversibility
Intelligence support	Low	Low	Medium	High
Rear area logistical support	High	Medium	Medium	Medium
Basing and transit of allied ground forces	High	Medium	Medium	Low
Overflights	Medium	Low	Medium	High
Ground troops in conflict zone	High	High	High	Medium

[25] Other research has found that ease of withdrawing military forces, which we term *reversibility*, was a characteristic that allies considered in deciding to contribute to coalition operations against Iraq in the Persian Gulf War (Bennett, Lepgold, and Unger, 1994).

another ally. Importantly, we explain the logic that motivates decisionmakers to consider each factor. The 13 factors represent a synthesis of the full range of factors we identified in the literature. The benefit of this holistic approach to identifying insights from the international relations literature is that it allows us to identify factors that might not be part of the current debate about a Baltic scenario.

This framework provides indicators that analysts and planners can monitor to assess changes in allied willingness to make military contributions in any scenario. Because we draw from theory and historical evidence in a range of cases that might not be directly applicable to every contingency, we cannot predict a specific ally's contribution in a specific situation. Rather, the framework offers policymakers and planners a systematic way to identify the factors that might shape each ally's decisions to contribute. By understanding these central factors, policymakers and planners can more effectively tailor strategies for addressing each ally's concerns, increasing the likelihood that they will make a military contribution.

The second part of the analysis, presented in Chapter Three, considers the relative importance of each of these factors in unconventional and conventional scenarios. In both cases, we consider how the characteristics of each scenario—such as the ability to attribute antiregime activities in the Baltics to Russia—might affect the importance of each factor in allied decisionmaking. We use a Baltic scenario to make the discussion more concrete, but the same analysis would apply to any conventional or unconventional Russian attack on a NATO member.

We then used this framework to identify tools that Russia might use to influence allies' decisions about military responses (Chapter Five) and steps that the United States can take to reduce allies' vulnerability to manipulation and promote a unified response to an attack on a NATO member (Chapter Six).

CHAPTER TWO

Factors Affecting an Ally's Decision to Provide Military Support

In this chapter, we outline factors that could affect whether and how a state contributes to collective defense through NATO or a coalition of the willing. We group these factors into three broad categories: domestic politics, perceptions of Russia, and alliance politics.[1] Throughout, we provide illustrative examples of these mechanisms at work in the decisionmaking of NATO members.

Domestic Politics

Many analysts have raised concerns about NATO's willingness to respond to a Russian attack because of the findings of public opinion polls in Europe. In 2017, the Pew Foundation found that a sizeable proportion of the public in some NATO countries does not support upholding Article 5 commitments. A 2017 Pew Foundation poll found that 60 percent of German, 55 percent of British, and 54 percent of Spanish respondents opposed using military force to defend a NATO ally.[2] These findings were consistent with a 2016 Bertelsmann Foundation poll in Germany, which found that 57 percent of Germans responded in the negative when asked whether German soldiers should stand "in defense of NATO members Poland and the Baltic states if they are attacked by Russia."[3]

However, as is detailed in this section, research on domestic politics and foreign policy has shown that the connection between public opinion and national security decisionmaking is complex and context dependent.[4] Decisionmakers tend to be more sensitive to public opinion as elections approach and when elite perspectives are

---

[1] Davidson applies similar categories in his analysis of allied decisionmaking of U.S.-led interventions in Kosovo, Afghanistan, and Iraq (Davidson, 2011).

[2] Stokes, 2017.

[3] Gabriele Schöler, *Frayed Partnership: German Public Opinion on Russia*, Gütersloh, Germany, and Warsaw, Poland: Bertelsmann Stiftung and Institute of Public Affairs, Poland, April 2016.

[4] For an overview of this literature, see Matthew A. Baum and Philip B. K. Potter, "The Relationships Between Mass Media, Public Opinion, and Foreign Policy: Toward a Theoretical Synthesis," *Annual Review of Political Sci-*

opinion. For NATO members that already have deployments in the Baltics, an attack on NATO forces, if clearly attributable to Russia, could shift public attitudes in support of a military response.

Third, elites can influence public attitudes on foreign policy issues.[10] If elites across political lines support contributing to a NATO operation, they might be able to gain public support. A 2016 study of domestic support for NATO's International Security Assistance Force (ISAF) mission to Afghanistan found that plausible, coherent, and consistent narratives were at least partially effective in reducing public opposition in allied countries.[11] In Germany, Defense Minister Karl-Theodor zu Guttenberg's narrative surrounding the 2009 Kunduz air strike reportedly contributed to passive acceptance of Germany's role by an otherwise critical German public.[12] In another example, the German government appealed to core themes of German political culture to strengthen public support for German contributions to NATO's 1999 operations in Kosovo: aversion to state aggression (in this case, Serbian aggression), affinity for multilateral responses (in this case, NATO), and prioritization of human rights.[13]

Political elites could also reduce support for NATO operations. In the Netherlands, a robust 2010 campaign for withdrawal from Afghanistan by the Labor Party and opposition Freedom Party might have strengthened public skepticism about the mission; the resulting political crisis ultimately caused the 2010 collapse of the Dutch government.[14] As we will discuss in subsequent sections, the electoral rise of new political figures in NATO governments that promote populist, nationalist, and antiestablishment messages have challenged a traditional elite consensus on NATO. These voices could serve to exacerbate existing doubts within NATO publics about national contributions to an alliance mission.

---

[10] Michael Howard, *The Causes of War and Other Essays*, Cambridge, Mass.: Harvard University Press, 1983, p. 317.

[11] Lawrence Freedman, "The Possibilities and Limits of Strategic Narratives," in Beatrice de Graaf, George Dimitriu, and Jens Ringsmose, eds., *Strategic Narratives, Public Opinion, and War: Winning Domestic Support for the Afghan War*, New York: Routledge, 2015, p. 24. See also Beatrice de Graaf, George Dimitriu, and Jens Ringsmose, "Conclusion—How to Operate Strategic Narratives: Interweaving War, Politics, and the Public," in Beatrice de Graaf, George Dimitriu, and Jens Ringsmose, eds., *Strategic Narratives, Public Opinion, and War: Winning Domestic Support for the Afghan War*, New York: Routledge, 2015, p. 364.

[12] While there was no public outcry or protest, neither was there public approval: Polling data showed that in 2010, the general German public approval for combat in Afghanistan was only 44 percent (Robin Schroeder and Martin Zapfe, "'War-Like Circumstances'—Germany's Unforeseen Combat Mission in Afghanistan and Its Strategic Narratives," in Beatrice de Graaf, George Dimitriu, and Jens Ringsmose, eds., *Strategic Narratives, Public Opinion, and War: Winning Domestic Support for the Afghan War*, New York: Routledge, 2015, p. 188).

[13] Mark Wintz, *Transatlantic Diplomacy and the Use of Military Force in the Post–Cold War Era*, New York: Palgrave Macmillan, 2010, p. 111.

[14] de Graaf, Dimitriu, and Ringsmose, 2015, p. 358; see also Reed Stevenson and Aaron Gray-Block, "Dutch Government Falls over Afghan Troop Mission," *Reuters*, February 19, 2010.

Public views on the use of force are highly context dependent, which makes them difficult to predict in advance of a crisis. Moreover, it is elite perceptions of public opinion, which can differ from actual public opinion, that have been shown to determine outcomes.[15] Further complicating matters, the effect of public opinion on foreign policy choices can depend on a government's electoral vulnerability and the extent of elite consensus on foreign policy.

Electoral Vulnerability of the Government

An upcoming election can make leaders more sensitive to public opposition, especially if they are electorally vulnerable and the opposition is running on an alternative foreign policy agenda.[16] For example, public opposition to U.S. military action in advance of the November presidential election contributed to President Dwight D. Eisenhower's decision not to act militarily during the 1956 Suez Crisis.[17] Similarly, German Chancellor Gerhard Schroeder's decision to oppose the U.S. invasion of Iraq was influenced in part by his Social Democratic party's weak standing in the polls in the summer of 2002.[18] More recently, Germany's decision not to participate in NATO's 2011 Operation Unified Protector campaign in Libya might have been driven by concerns that participation would undermine the governing party's prospects in the spring 2012 Bundesrat elections.[19] Electoral punishment for unpopular wars can be decisive: In 2004, widespread discontent with the incumbent government's commitment of Spanish forces to the U.S.-led invasion of Iraq—as well as its handling of terrorist bomb attacks—prompted an unexpected parliamentary victory by Spain's antiwar Socialist party and led to the withdrawal of Spanish forces from Iraq.[20]

The sensitivity of elected officials to public opinion might also depend on the risk of defections from within their governing coalition. For example, French Prime Minister Guy Mollet was initially reluctant to use force during the 1956 Suez crisis despite considerable domestic pressure in favor of intervention, partly because his coalition partners threatened to defect if the intervention went poorly.[21] In a more recent example, coalition negotiations between Italian Prime Minister Silvio Berlusconi's ruling

---

[15] For an overview of this literature, see Aldrich et al., 2006, pp. 491–495.

[16] Davidson also addresses the interactions between public opinion, electoral proximity, and opposition stances in determining government calculations about political survival (Davidson, 2011).

[17] David P. Auerswald, "Inward Bound: Domestic Institutions and Military Conflicts," *International Organization*, Vol. 53, No. 3, Summer 1999.

[18] John Hooper, "German Leader Says No to Iraq War," *The Guardian*, August 5, 2002.

[19] The Bundesrat is one of the five constitutional bodies in Germany, which permits the federal states to participate in national legislation and administration (Sperling, 2016).

[20] Lizette Alvarez and Elaine Sciolino, "Bombings in Madrid: Election Outcome; Spain Grapples with Notion that Terrorism Trumped Democracy," *New York Times*, March 17, 2004.

[21] Auerswald, 1999, p. 485.

People of Freedom party and the Northern League were a significant factor in Italy's initial decision not to participate in Operation Unified Protector.[22]

Conversely, when there is elite consensus on support for a foreign military contribution, the influence of public opinion should be lower.[23] For example, despite low levels of public support for the deployment of troops to Afghanistan, many European countries contributed troops to ISAF. This was possible in part because of elite consensus about the importance of upholding NATO commitments. As one scholar explained, elite consensus in Germany, France, Canada, and Italy "had the effect of inoculating the leadership from electoral punishment. Since the opposition has a similar stand, leaders are thereby less concerned about being challenged or losing votes to competitors and can therefore continue the unpopular Afghanistan deployment."[24] Elite consensus during the course of a conflict can also serve to maintain public support for a war, even when casualties are high.[25]

The advent of significant antiestablishment minority parties that question fundamental assumptions about the NATO alliance has made elite consensus more tenuous. Germany's nationalist Alternative für Deutschland (AfD) party, which won 90 seats (12.6 percent) in the 2016 German parliamentary elections, includes in its party platform a call for the withdrawal of all allied troops from German soil.[26] France's Front National candidate Marie Le Pen, who made it to the final round of the 2017 French presidential election but ultimately only garnered only about a third of the popular vote, has long called for France's withdrawal from NATO.[27] During a conflict with Russia, these voices and others could seek to apply political pressure to governing leaders to minimize national contributions to a NATO operation. If a significant election was on the horizon—and if public opinion was opposed to conflict—the pressure would be particularly acute.

---

[22] Sperling, 2016, p. 71.

[23] Elizabeth N. Saunders, "War and the Inner Circle: Democratic Elites and the Politics of Using Force," *Security Studies*, Vol. 24, No. 3, September 2015; Janne Haaland Matláry and Magnus Petersson, "Introduction: Will Europe Lead in NATO?" in Janne Haaland Matláry and Magnus Petersson, eds., *NATO's European Allies: Military Capability and Political Will*, New York: Palgrave Macmillan, 2013, p. 3.

[24] Sarah Kreps, "Elite Consensus as a Determinant of Alliance Cohesion: Why Public Opinion Hardly Matters for NATO-Led Operations in Afghanistan," *Foreign Policy Analysis*, Vol. 6, No. 3, July 2010.

[25] Adam J. Berinsky, "Assuming the Costs of War: Events, Elites, and American Public Support for Military Conflict," *Journal of Politics*, Vol. 69, No. 4, November 2007; Saunders, 2015. As we discuss later in this report, high levels of external threats can strengthen elite consensus and diminish the influence of public opinion (Kreps, 2010, pp. 202–203).

[26] AfD, *Manifesto for Germany: The Political Programme of the Alternative for Germany*, undated.

[27] Stephanie Pezard, "The Front National and the Future of French Foreign Policy," *War on the Rocks*, March 25, 2015.

Foreign Policy Decisionmaking Structure

States with fewer *veto players* in their foreign policy decisionmaking structure are more likely to join a military response and to make more-significant contributions. Veto players are the people or institutions that have to assent for a policy. The more veto players, the more difficult it is to change policy and the more likely that any changes will be modest or involve significant compromises.[28] Among allies that contributed troops to the NATO operation in Afghanistan, those with more-inclusive foreign policy decisionmaking placed more restrictions on how those forces were used than allies with more-centralized processes.[29] Germany, for example, has a constitutional requirement that the legislature approve troop commitments; while Germany has made significant troop contributions, it also has placed strict caveats on its forces.[30]

This research suggests that in a crisis with Russia, NATO countries with more veto players, such as those with coalition governments, are less likely to contribute and, if they do, will do so more conditionally. Other allies, especially those with presidential or majoritarian parliamentary systems, have less-restrictive, more-centralized decisionmaking processes. In France, the president need only inform the French National Assembly and Senate of a decision to deploy French troops.[31] All things being equal, these states should be more likely to contribute and to do so with fewer restrictions.

Perceptions of Russian Threat and Vulnerability to Russian Retaliation

An allied government's perception of the threat Russia poses and the risk of Russian retaliation are likely to affect its willingness to participate in coalition or NATO actions in the Baltics. While the previous section discussed public opinion, this section focuses on decisionmaker perspectives. Perceptions of Russia can produce countervailing considerations for decisionmakers, as summarized in Table 2.2 and detailed below. Perceptions of Russia as opportunistic and ambitious or as the state's primary national security threat will tend to increase consideration of more-significant military contri-

[28] The general logic of veto players and policy change is outlined in George Tsebelis, "Decision Making in Political Systems: Veto Players in Presidentialism, Parliamentarism, Multicameralism and Multipartyism," *British Journal of Political Science*, Vol. 25, No. 3, July 1995, p. 289. For a brief overview of the literature in the context of foreign policy, see Stephen M. Saideman and David P. Auerswald, "Comparing Caveats: Understanding the Sources of National Restrictions Upon NATO's Mission in Afghanistan," *International Studies Quarterly*, Vol. 56, No. 1, March 2012, pp. 70–71.

[29] Saideman and Auerswald, 2012.

[30] In Germany, opponents might also mount a legal challenge in the Federal Constitutional Court for any element of an operation falling outside parliamentary mandate (Sperling, 2016, pp. 70–71). The German military imposes additional constraints on itself in anticipation of parliamentary concern (Gale A. Mattox, "Germany: The Legacy of the War in Afghanistan," in Gale A. Mattox and Stephen M. Grenier, eds., *Coalition Challenges in Afghanistan: The Politics of Alliance*, Stanford, Calif.: Stanford University Press, 2015, p. 94).

[31] Sperling, 2016, pp. 70–71; Saideman and Auerswald, 2012, p. 72.

wake of the 2008 Russo-Georgian war, a group of 22 prominent central European intellectuals and former leaders argued that "Russia is back as a revisionist power" and called for contingency planning for Russian attacks and prepositioning military equipment in the region.[34] In June 2018, the "Bucharest Nine" (eight presidents and one speaker of parliament from central and eastern Europe) publicly pledged closer military cooperation in response to an "aggressive" Russia's threats against a free, peaceful Europe.[35] These viewpoints stand in stark contrast to those who see Russian aims as limited. Former German Social Democratic Party (SPD) Chancellor Gerhard Schroeder argued that Russia had significant domestic challenges and "that no one in the Moscow leadership has an interest in military conflicts."[36]

The motivations that each allied government sees behind past Russian behavior, including its aggression in Georgia and in Ukraine, could also affect that ally's response. If decisionmakers believe Russian interventions were motivated by opportunism, they are more likely to contribute to a military response to signal NATO's resolve and deter further Russian aggression.[37] Conversely, those who believe Russia is motivated by legitimate security concerns are likely to be more concerned that a NATO military response in the Baltic region will only exacerbate Russian insecurity and provoke unnecessary conflict.[38]

There is evidence that some within NATO view recent Russian aggression as a security-motivated response to past U.S. and NATO policies, such as the Libya operation, NATO missile defense, and NATO enlargement.[39] This perspective has been expressed within major European political parties, including Germany's SPD. In June 2016, after a major NATO exercise in Poland, SPD member and German Foreign

National Security Strategy and Strategic Defence and Security Review 2015: A Secure and Prosperous United Kingdom, London: Her Majesty's Stationery Office, CM9161, November 2015).

[34] Valdas Adamkus, Martin Butora, Emil Constantinescu, Pavol Demes, Lubos Dubrovsky, Matyas Eorsi, Istvan Gyarmati, Vaclav Havel, Rastislav Kacer, Sandra Kalniete, Karel Schwarzenberg, Michal Kovac, Ivan Krastev, Alexander Kwasniewski, Matt Laar, Kadri Liik, Janos Martonyi, Janusz Onyszkiewicz, Adam Rotfield, Vaira Vike-Freiberga, Alexandr Vondra, and Lech Walesa, "An Open Letter to the Obama Administration from Central and Eastern Europe," *Radio Free Europe/Radio Liberty*, July 16, 2009.

[35] "NATO Eastern Flank Members Pledge Closer Ties, Citing Russia," *Associated Press*, June 8, 2018.

[36] Cited in Stephen F. Szabo, *Germany, Russia, and the Rise of Geo-Economics*, New York: Bloomsbury Publishing, 2015, p. 46.

Table 2.2
Perceptions of the Russian Threat and Decisions to Support a Military Response

	More Likely to Contribute to a Military Response	Less Likely to Contribute to a Military Response
Perception of Russian aims and motivations	• Russia is ambitious and opportunistic	• Russian aims are limited or motivated by security concerns
Competing national security demands	• Homeland and vital interests face no other significant threats • Significant concerns exist about becoming a target of future Russian aggression	• Homeland or vital interests face other significant threats • Minimal concerns exist about becoming a target of future Russian aggression
Vulnerability to Russian military retaliation	• Proximity to Russia or lower military capability makes state more vulnerable	• More distance from Russia and more military capability make state less vulnerable
Vulnerability to Russian economic retaliation	• Diverse economy creates minimal reliance on Russia for energy or trade • Negative expectations exist about future economic relations with Russia	• More reliance on Russia exists for energy or trade • Positive expectations exist about future economic relations with Russia
Perception of escalation risk	• Low risk	• High risk

butions with greater military utility. However, concerns about Russian retaliation and escalation might lead states to consider smaller or less observable contributions.

Perception of Russian Aims and Motivations

Each allied government's view about Russian aims and motivations will affect its willingness to adopt a military response.[32] A NATO ally that believes Russian aims are limited (e.g., a finite territorial expansion against a non-NATO member) might be less likely to participate in a military response than an ally that believes Russia has plans to upend the entire European security order.

In recent years, public voices within NATO member states have varied in their views of Russian ambition. Those at one end of the spectrum see Russia as a serious threat and call for military responses. For example, in early 2015, the United Kingdom's senior-most NATO military officer warned that Russia's expansionist ambitions could become "an obvious existential threat to our whole being."[33] Similarly, in the

[32] For a general discussion of how views about an adversary's motivation lead to different policy prescriptions, see Robert Jervis, *Perception and Misperception in International Politics*, Princeton, N.J.: Princeton University Press, 1976.

[33] Peter Walker, "Russian Expansionism May Pose Existential Threat, Says NATO General," *The Guardian*, February 20, 2015. This view is consistent with the United Kingdom's 2015 Strategic Defense and Security Review, which identified Russian behavior at the top of the list of resurgent state-based threats (David Cameron,

Minister Frank-Walter Steinmeier argued that "what we shouldn't do now is inflame the situation further with loud saber-rattling and warmongering."[40] His deputy, Gernot Erler, warned that NATO military actions can "lead to uncontrollable situations, all the way up to war."[41]

In some cases, internal political balancing can influence the lens through which an allied government assesses and characterizes Russian behavior. Within Germany, the chancellery of Angela Merkel, a member of the Christian Democratic Union, has taken a more critical interpretation of Russian military activity in Ukraine than has the Ministry of Foreign Affairs, dominated by the SPD.[42] In Bulgaria, a national threat assessment identifying Russia as a threat to regional and European security failed to gain parliamentary approval because of resistance by the pro-Russia Bulgarian Socialist party, which had made electoral gains in the 2017 legislative elections. The follow-on report that was ultimately approved contained no references to Russia.[43] Similarly, although some Czech government officials have cited Russian disinformation efforts as "the greatest threat Europe has been facing since the 1930s," the Czech Republic formally has "no definitive stance about whether Russia is a threat to national security" because of divergences in perception between its government and pro-Russian President Milos Zeman.[44]

While this section primarily addresses the threat perceptions of decisionmakers, European publics that perceive Russia as a threat are also generally more likely to support a military response. A plurality of those polled by the Pew Foundation in every NATO country (except Germany and Italy) explicitly blamed Russia for the Ukraine crisis, including a majority of Polish respondents (57 percent), 44 percent of respondents in France, and 40 percent in the United Kingdom. About half or more respondents polled in most NATO nations agreed that Russia constitutes a major military threat to neighboring countries in addition to Ukraine: The highest levels of public concern were in Poland, at 70 percent, and the lowest were in Germany and Italy, at

[40] Matthew Karnitschnig, "NATO's Germany Problem," *Politico*, July 8, 2016.

[41] Karnitschnig, 2016.

[42] Following Russia's invasion of Crimea, Chancellor Merkel took a personal leading role in the establishment of transatlantic sanctions against Russia. In the immediate wake of the event, she told Christian Democratic Union colleagues that the invasion violated the principles of postwar order in Europe and that "what has hap-

38 and 44 percent, respectively.[45] These views appear to correlate with a public willingness to use force against Russia in defense of a NATO ally on its border: Italians and Germans polled had the lowest levels of support for such an operation, with Poland topping the list after the United States, Canada, and the United Kingdom.[46]

Competing National Security Demands

NATO countries have limited resources for defense and so must consider how the Russian threat stacks up against other threats. Many of these states are involved in other operations, such as training security forces in Afghanistan, responding to the Syrian refugee crisis, and combating terrorism abroad. As allies consider how to prioritize the Russian threat, leaders will have to think both about their short-term resources and their ability to regenerate forces given fiscal constraints, as well as industrial base issues in the medium- and long-term.

Allies that are militarily weak and proximate to Russia might be particularly likely to prioritize a Russian threat over other national security concerns.[47] For such countries as France, which has suffered a series of terrorist attacks since late 2014, low-level Russian aggression in the Baltic states might be deemed a relatively lower national security priority. Overstretched by counterterrorism missions in the Sahel, Middle East, and domestically, France has had to deprioritize other security initiatives, including contributions to NATO's military presence in eastern Europe.[48] Similarly, Italy has prioritized "Euro-Mediterranean" security in its military spending, suggesting that the Middle East and North Africa are currently a higher priority than Russian threats to the Baltics.[49] A similar dynamic has played out in previous conflicts, including Spanish contributions to the U.S.-led 2003 war in Iraq. Spain initially sent troops as part of the U.S.-led coalition.[50] However after domestic terrorist attacks by al-Qaeda–affiliated

---

[45] Katie Simmons, Bruce Stokes, and Jacob Poushter, *NATO Publics Blame Russia for Ukrainian Crisis, but Reluctant to Provide Military Aid*, Washington, D.C.: Pew Research Center, June 10, 2015, pp. 5, 17.

[46] Simmons, Stokes, and Poushter, 2015, pp. 5, 17.

[47] On proximity as a key determinant of threat assessments, see Stephen M. Walt, *The Origins of Alliances*, Ithaca, N.Y.: Cornell University Press, 1987. Territorial contiguity is a key predictor of interstate conflict, including war, suggesting that states are right to be most concerned about threats on their periphery (Stuart A. Bremer, "Dangerous Dyads: Conditions Affecting the Likelihood of Interstate War, 1816–1965," *Journal of Conflict Resolution*, Vol. 36, No. 2, June, 1992; John A. Vasquez, *The War Puzzle*, Cambridge: Cambridge University Press, 1993, p. 123).

[48] "Chapter Four: Europe," *Military Balance*, Vol. 117, No. 1, 2017, p. 78.

[49] On Italy's prioritization of security in the Mediterranean, see Alessandro Marrone and Michele Nones, *Italy and Security in the Mediterranean*, Rome: Instituto Affari Internazionali, 2016.

[50] Philip H. Gordon and Jeremy Shapiro, *Allies at War: America, Europe, and the Crisis over Iraq*, Washington, D.C.: Brookings Institution Press, 2004, pp. 127, 144.

express displeasure with plans for an enhanced NATO presence in the region.[61] However, while both the Polish and the Lithuanian governments have expressed grave concern about the development and denounced the efforts at coercion, neither has backed away from their strong support for NATO's regional buildup. Though concerns about nuclear use will be greatest at higher levels of conflict, allies might consider the risk of a wider conventional or nuclear war even in an unconventional scenario. They might consider key questions, such as: Can a conflict in the Baltics stay limited? What is Russia's threshold for nuclear use? Are Russia's nuclear threats credible?

Alliance Politics

A state's assessment of the cost and benefits of a military response could also be affected by considerations that are specific to the alliance itself. In general, two broad concerns tend to pervade alliance politics: the risk of dissolution or abandonment (that the alliance itself will fall apart or that key allies will withdraw their commitments) and the risk of *entrapment* (the possibility of being dragged into a conflict unnecessarily).[62] A state's decision to participate in a NATO response could be affected by how it balances these concerns.

Any scenario that presumes an informal coalition of the willing rather than a formal NATO response complicates calculations about the costs and risks of dissolution or abandonment. On the one hand, even in the absence of a formal NATO mandate, some allies might see a lack of coalition participation as a betrayal of NATO commitments and decide to punish nonparticipants in some way. This would be most likely in a case of a conventional Russian attack on a NATO member. In a more ambiguous scenario, some allies might have strong opposition to the operation and seek to punish those that operate outside of NATO auspices. Countries could therefore plausibly face countervailing pressures from allies.

Table 2.3 summarizes how five categories of alliance considerations, detailed in the remainder of this section, might affect a NATO ally's willingness to support a military response in a Baltic scenario.

Participation of Other Allies

As an ally weighs a potential contribution to collective defense, it is likely to consider who else is participating.[63] Research on alliance burden-sharing has focused on a mix

---

[61] Dmitry Solovyov and Andrius Sytas, "Russia Moves Nuclear-Capable Missiles into Kaliningrad," *Reuters*, October 8, 2016.

[62] Glenn H. Snyder, "The Security Dilemma in Alliance Politics," *World Politics*, Vol. 36, No. 4, July 1984.

[63] This factor was highlighted during a series of RAND tabletop exercises in May–June 2018 for another RAND project.

Table 2.3
Alliance Politics and Decisions to Support a Military Response

	More Likely to Adopt a Military Response	Less Likely to Adopt a Military Response
Participation of other allies	• Militarily powerful allies are involved, making success more likely • Many allies are participating, making it more legitimate	• Militarily powerful allies are not participating • Few allies are participating, making the intervention less legitimate
Alignment of goals among participants	• State's goals for the operation are aligned, especially with leading allies • Minimal concerns about provocations by frontline states	• State has much less ambitious goals than leading allies • State has concerns about provocation by frontline states
Ability to restrain coalition members	• State expects allies will be restrained within coalition decisionmaking structure	• State believes some allies will ignore coalition concerns or operate unilaterally
Consequences of alliance dissolution or abandonment	• State is reliant on NATO for security and expects punishment for non-participation • State deeply values NATO for other non-security reasons	• State is not reliant on NATO for security or does not expect NATO punishment for nonparticipation • State does not deeply value NATO
Punishment by noncoalition allies	• Nonparticipants are few or plan to stay neutral	• Many nonparticipants that intend to punish defectors

of rational calculations and the normative considerations that states make. Some analysts have applied collective action theory to explain why, during peacetime, smaller NATO allies are likely to minimize contributions or freeride. If these allies believe that others will provide collective defense—even if they do not contribute—they have little to gain by burden-sharing. However, this dynamic would be less applicable in the event of a conventional Russian attack for two reasons. First, as detailed below, states that do not contribute could be punished more harshly by other allies for failing to respond to a Russian attack than they are in peacetime. Second, each ally's contributions could materially affect the outcome if the crisis were to escalate to a wider war.[64] Put another way, in the face of an overwhelming common security threat—direct conflict with Russia—a freeriding approach could potentially be more costly for NATO members than burden-sharing.[65]

[64] Mancur Olson, Jr., and Richard Zeckhauser, "An Economic Theory of Alliances," *Review of Economics and Statistics*, Vol. 48, No. 3, August 1966, p. 278.

[65] David A. Lake, *Entangling Relations: American Foreign Policy in Its Century*, Princeton, N.J.: Princeton University Press, 1999; Michael A. Allen, Julie van Dusky-Allen, and Michael E. Flynn, "The Localized and Spatial

Allies will also consider whether the operation has enough military support to be successful. Whether the United States—the strongest member of the alliance—contributes could be critical to the coalition's prospects for success. If the United States is making significant contributions, an allied government might feel more confident in the operation and, therefore, be more willing to contribute.

State decisions are not affected just by rational calculations about likely consequences; they are also determined by decisionmaking heuristics and the state's view of its role in the world. [66] Some allied pairs develop the tendency to look to one another for decisions because of "embedded habits of cooperation" borne of repeated security collaborations.[67] Such habits might be particularly pronounced in those allies that have combined forces, such as the French-German military brigade. Moreover, states might be more likely to contribute if they perceive a military operation to be legitimate and consistent with their identity, which is determined in part by the participation of other allies.[68]

Alignment of Goals Among Participants

As described above, a state might worry about the risk of entrapment—being dragged into a conflict by its alliance commitments. Historically, U.S. allies, such as Japan, have had such concerns and have taken steps in peacetime to reduce such risks, such as ensuring that formal agreements contain explicit statements about the scope of alliance commitments.[69]

For a state considering joining a coalition in the midst of a crisis, an analogous concern might arise. A state might worry that some allies have wider aims at the outset of the conflict or might develop wider ambitions once a military response begins.[70] Entrapment can occur when one of the coalition members takes a provocative action and other coalition members feel compelled to defend the provocateur in order to defend the credibility or cohesion of the coalition.[71] For example, if a Baltic state took

---

Effects of U.S. Troop Deployments on Host-State Defense Spending," *Foreign Policy Analysis*, Vol. 12, No. 4, October 2016.

[66] James G. March and Johan P. Olsen, "The Institutional Dynamics of International Political Orders," *International Organization*, Vol. 52, No. 4, Autumn 1998.

[67] Ruike Xu, *Alliance Persistence Within the Anglo-American Special Relationship: The Post–Cold War Era*, Cham, Switzerland: Springer International Publishing, 2017.

[68] For an overview of the literature on the logic of appropriateness in state decisions about the use of force, see Sarah Kreps, "When Does the Mission Determine the Coalition? The Logic of Multilateral Intervention and the Case of Afghanistan," *Security Studies*, Vol. 17, No. 3, September 2008.

[69] Tongfi Kim, "Why Alliances Entangle but Seldom Entrap States," *Security Studies*, Vol. 20, No. 3, August 2011.

[70] For a general discussion of the expansion of war aims, see Labs, 1997.

[71] Alternatively, a state might become a target of Russian retaliation for an action it did not support in the first place and subsequently feel compelled to respond in self-defense. For an overview of the literature on the risk of

steps to repress their ethnic Russian populations at the outset of a crisis, Russia might escalate the conflict by sending in special operations forces to defend these populations. In this situation, allies might still feel compelled to come to the Baltic state's aid even though it had adopted a provocative policy without the consent of other allies. A participant might also worry that any misalignment in goals will make it difficult to come to a negotiated settlement with Russia to end the conflict. In other words, if the coalition is divided about its aims, the crisis or war could last much longer than it would if goals were aligned.[72]

NATO's endurance and members' willingness to continue to pay peacetime costs of exercises and military coordination indicate shared interests.[73] However, substantial divergences have continued and, in some cases, have affected allies' willingness to contribute to military operations both within and beyond the NATO context. For example, while NATO invoked Article 5 after the September 11 terrorist attacks, significant divergences appeared in the years that followed. When the United States sought to a establish a "coalition of the willing" outside of NATO to invade Iraq, key allies refused to participate and condemned U.S. actions.[74] More recently, Baltic leaders have publicly supported the use of lethal force in response to Russian unconventional incursions into their territory, but other allies have been less willing to commit to a military response.[75] Russian leaders have long believed that the United States seeks regime change.[76] Allies that share that view might worry that a U.S.-led, non-NATO coalition of the willing against Russia would pursue those more expansive aims once fighting begins.

To the extent that an ally believes that its goals are aligned with other coalition members, especially frontline states or the coalition leader, it is more likely to participate. However, allies that perceive a misalignment of goals or believe that the coalition leader could develop more-expansive aims as fighting continues will be more cautious about joining. These concerns could lead to three outcomes. First, if these concerns are acute, the ally might decide not to participate in any form. Second, an ally might offer political or financial support or make military contributions that are more easily

entrapment in alliances, see Michael Beckley, "The Myth of Entangling Alliances: Reassessing the Security Risks of U.S. Defense Pacts," *International Security*, Vol. 39, No. 4, Spring 2015; Kim, 2011.

72 Patricia A. Weitsman, "Intimate Enemies: The Politics of Peacetime Alliances," *Security Studies*, Vol. 7, No. 1, Autumn 1997; Patricia A. Weitsman, "Alliance Cohesion and Coalition Warfare: The Central Powers and Triple Entente," *Security Studies*, Vol. 12, No. 3, Spring 2003, p. 86; Snyder, 1984, p. 474; Thomas J. Christensen, *Worse Than a Monolith: Alliance Politics and Problems of Coercive Diplomacy in Asia*, Princeton, N.J.: Princeton University Press, 2011.

73 James D. Morrow, "Alliances: Why Write Them Down?" *Annual Review of Political Science*, Vol. 3, June 2000, pp. 68–70.

74 Weitsman, 2003, p. 112.

75 See Sam Jones, "Estonia Ready to Deal with Russia's 'Little Green Men,'" *Financial Times*, May 13, 2015.

76 Defense Intelligence Agency, *Russia Military Power: Building a Military to Support Great Power Aspirations*, Washington, D.C., DIA-11-1704-161, 2017, pp. 15–16.

reversed (e.g., deploying naval assets instead of ground troops) if a coalition member adopts more-expansive aims.[77] Finally, a state might evaluate whether allies with more-expansive aims could be restrained through existing or new decisionmaking structures, as detailed in the next section.

Ability to Restrain Coalition Members

Concerns about a mismatch of goals could be overcome if coalition participants are more confident in their ability to influence the coalition's actions. Potential participants could have concerns about restraining any ally, but in a Baltic scenario, these concerns are likely to be most acute with respect to the most powerful and leading members of the coalition and the Baltic states.

Typically, weaker states are thought to have less leverage in the alliance and, therefore, greater concerns about entrapment by a stronger ally like the United States.[78] However, weaker allies do have some ways to restrain more-powerful allies. First, long-standing diplomatic and military relationships among NATO members mean that states might have insight into their allies' decisionmaking processes and the contacts needed to influence those processes.[79] Second, allies also have some sources of leverage over a coalition leader, including the ability to withdraw participation or place limits on access to military bases or airspace.[80] Such actions could restrain a coalition leader's behavior by affecting its ability to carry out operations, raising the costs of a military action, or undermining the domestic or international legitimacy of the operation. To the extent that a coalition leader values the support of its allies in the longer term, threats of exit from the coalition or NATO itself could affect a coalition leader's decisionmaking.[81]

There is evidence that NATO allies used NATO decisionmaking structures to restrain U.S. actions during the air war over Kosovo. Wesley Clark, who was NATO's

---

[77] Bennett, Lepgold, and Unger, 1994, p. 44. States might also seek to make military contributions to support another operation or line of effort, as Germany did during NATO's Operation Unified Protector by sending AWACS crews to Afghanistan.

[78] James D. Morrow, "Alliances and Asymmetry: An Alternative to the Capability Aggregation Model of Alliances," *American Journal of Political Science*, Vol. 35, No. 4, November 1991; Jeremy Pressman, *Warring Friends: Alliance Restraint in International Politics*, Ithaca, N.Y.: Cornell University Press, 2008; Gene Gerzhoy, "Alliance Coercion and Nuclear Restraint: How the United States Thwarted West Germany's Nuclear Ambitions," *International Security*, Vol. 39, No. 4, Spring 2015.

[79] G. John Ikenberry, *After Victory: Institutions, Strategic Restraint, and the Rebuilding of Order After Major Wars*, Princeton, N.J.: Princeton University Press, 2001.

[80] Michael J. Lostumbo, Michael J. McNerney, Eric Peltz, Derek Eaton, David R. Frelinger, Victoria A. Greenfield, John Halliday, Patrick Mills, Bruce R. Nardulli, Stacie L. Pettyjohn, Jerry M. Sollinger, and Stephen M. Worman, *Overseas Basing of U.S. Military Forces: An Assessment of Relative Costs and Strategic Benefits*, Santa Monica, Calif.: RAND Corporation, RR-201-OSD, 2013, p. xxiii.

[81] Songying Fang, Jesse C. Johnson, and Brett Ashley Leeds, "To Concede or to Resist? The Restraining Effect of Military Alliances," *International Organization*, Vol. 68, No. 4, Fall 2014.

Supreme Allied Commander, complained about the influence of NATO members on military decisions, including campaign phasing and individual targeting.[82] In a now-famous incident at the Pristina airport, the British commander of NATO's Allied Rapid Reaction Corps was able to leverage his chain of command in London to thwart Clark's plan to block the airport runway with helicopters—a move that the British commander believed could have triggered a security incident between NATO and Russia.[83] Ultimately, concerns about constraints posed by NATO structures in Kosovo contributed to the U.S. decision to form a coalition of the willing for initial operations in Afghanistan in 2001.[84]

NATO members that do not participate could still face direct Russian retaliation or other consequences. Given that possibility, members who do not entirely support the aims of the operation might still participate in order to seek greater influence over operational decisions than they would have if they were on the outside.[85] This logic appears to have influenced France's decision to join NATO's air war over Kosovo. French Secretary General for Defense and National Security Louis Gautier argued that French participation in the operation meant that France was "able in Kosovo to impose itself in the decisionmaking process and affect strategy."[86]

Consequences of Alliance Dissolution or Abandonment

Allies might worry not only about the risk of entrapment by their allies, but also of the risk of abandonment.[87] While allied governments might not always calculate contributions to a NATO mission to be in their immediate self-interest, they might believe that participation increases likelihood of allied support should *they* one day be attacked. More generally, a NATO member government might provide military support to a NATO operation because of concerns that failure to do so would jeopardize the strength and credibility of the alliance as a whole. NATO allies have not traditionally wielded coercive threats of abandonment or other punishment to foster allied participation. However, July 2018 warnings from the U.S. President that the United States

---

[82] Dana Priest, "United NATO Front Was Divided Within," *Washington Post*, September 21, 1999.

[83] Mike Jackson, "Gen Sir Mike Jackson: My Clash with NATO Chief," *The Telegraph*, September 4, 2007.

[84] Suzanne Daley, "After the Attacks: The Alliance; For First Time, NATO Invokes Joint Defense Pact with U.S.," *New York Times*, September 13, 2001.

[85] On the general phenomenon of weaker states joining institutions to restrain stronger states, see Ikenberry, 2001.

[86] Scholars also have pointed to "the risk that Paris might find itself marginalized in the main debates over the necessary transformation of the post–Cold War Alliance" as one motivation for France's 2007 decision to rejoin NATO military command (Frédéric Bozo, "Explaining France's NATO 'Normalisation' Under Nicolas Sarkozy [2007–2012]," *Journal of Transatlantic Studies*, Vol. 12, No. 4, December 2014).

[87] For an assessment of how the threat of punishment for defecting from an institutional commitment can affect state calculations, see Robert O. Keohane, *After Hegemony: Cooperation and Discord in the World Political Economy*, Princeton, N.J.: Princeton University Press, 1984, pp. 99–100.

might "go [its] own way" in the absence of increased allied defense spending could reasonably have raised concerns among allies about U.S. political leadership's commitment to their security.[88]

Weaker allies in close proximity to Russia have the most to lose from allied abandonment. These states could be the next target if Russia were to become more ambitious. Therefore, they have incentives to participate in a collective military response.[89] For such frontline countries as Poland, the consequences of allied abandonment during a military confrontation with Russia would be devastating. Polish security studies scholar Marek Pietras argued that Polish policy makers undertook the decision to contribute over 2,500 troops and civilian personnel to the ISAF mission in Afghanistan, despite strong domestic opposition, because it "provided grounds for expecting reciprocity and solidarity on the part of other member states under Article 5."[90] Even stronger states, such as Germany, have cited NATO and alliance solidarity as a central part of a national security strategy and have had concerns that nonparticipation could have far-reaching implications.[91]

NATO members might worry about alliance collapse or the exit of powerful allies, such as the United States, for reasons beyond homeland defense. Either could compromise NATO's ability to conduct other operations, including humanitarian relief, counterterrorism, and enforcement of international law. For European security elites who have incorporated NATO into their core national security cultures, the existence of NATO and its associated organizations has significance beyond pure rational calculation. Although the United Kingdom might rely heavily on alliance security guarantees, behavioral and normative factors might also lead the United Kingdom to prioritize the U.S.-British "special relationship" and NATO, leading it to contribute to alliance operations in Afghanistan and elsewhere.[92] Germany also values NATO

---

[88] David M. Herszenhorn and Lili Bayer, "Trump's Whiplash NATO Summit," *Politico*, July 12, 2018.

[89] On alliance abandonment, see Snyder, 1984, p. 466; Morrow, 1991. In theory, if a state had an alternative alliance option, the consequences of abandonment would be lower. Davidson similarly notes that weaker states are more likely to have high value for large allies than stronger states, and that states are more likely to value large allies when they face a large, menacing threat (Davidson, 2011).

[90] Between 2007 and 2010, Polish public opinion polls showed between 75 and 80 percent of the population disapproved of Poland sending troops to Afghanistan (Marek Pietras, "Poland's Participation in NATO Operations," in Janne Haaland Matláry and Magnus Petersson, eds., *NATO's European Allies: Military Capacity and Political Will*, New York: Palgrave Macmillan, 2013, p. 211).

[91] Federal Ministry of Defence, *White Paper 2016: On German Security Policy and the Future of the Bundeswehr*, Berlin: Federal Ministry of Defence, Federal Government of Germany, July 13, 2016.

[92] Phillip Walter Wellman, "UK to Nearly Double Troops in Afghanistan After Trump Request," *Stars and Stripes*, July 11, 2018. In addition to "mutual reciprocity" and "recurrent common threats," one 2017 assessment of the U.S.-United Kingdom alliance attributes its enduring strength to such behavioral factors as "embedded habits of cooperation," as well as the more normative "congenial (like-minded) partnership" (Xu, 2017).

for a mix of rationalist and cultural reasons.[93] Therefore, while Germany chose not to participate in the NATO campaign in Libya directly, it indirectly supported the effort by sending by sending German NATO AWACS crews to Afghanistan, freeing other NATO crews for operations in Libya. The German Ministry of Defense described the move as a "political sign of our solidarity with the alliance."[94] Ultimately, states that place higher value on the "public goods" provided by the existence of the NATO alliance might be willing to bear a share of the burden during a military campaign.[95]

The extent to which a NATO member government might value the perpetuation of the NATO alliance is not immutable.[96] When nationalist, populist, and antiestablishment parties are voted into decisionmaking positions, their national security policies do not typically reflect concern about the potential for NATO alliance dissolution or abandonment. Italy's Five Star movement called for Italy to play a reduced role in NATO during the 2018 Italian parliamentary elections.[97] Within weeks of its ascent into the Italian government, the party had blocked moves to increase defense spending to the 2 percent of gross domestic product required by NATO and had begun to "reevaluate [Italy's] presence in international missions from a point of view of their effective relevance to the national interest."[98]

Given the length of the U.S. commitment to NATO and NATO's central role in U.S. grand strategy since World War II, a U.S. exit might, at times, have seemed remote.[99] However, several factors could make some allies see it as a possibility. First, because of overwhelming U.S. power advantages, its alliances are asymmetric: U.S. allies are more dependent on the United States for security than vice versa. This means that the United States could see changing the status of its alliance commitments as less costly than a weaker ally might. Second, as noted in Chapter One, changing domestic politics in the United States have led to calls from U.S. senior leadership for making

[93] For example, the 2016 German white paper linked NATO to the preservation of international law and human rights (Federal Ministry of Defence, 2016).

[94] "Germany's Libya Contribution: Merkel Approves AWACS for Afghanistan," *Spiegel Online*, March 23, 2011. See also Sperling, 2016, p. 78.

[95] On burden-sharing in alliances generally, see Olson and Zeckhauser, 1966.

[96] Davidson also highlights the significance of alliance value in his analysis of ally decisionmaking, and notes that state value for alliance relationships might vary depending on the sitting government (Davidson, 2011).

[97] James McBride, "What to Know About Italy's 2018 Elections," Council on Foreign Relations, February 14, 2018.

[98] Tom Kington, "Future of F-35 in Italy Remains a Mystery Under New Government," *DefenseNews*, May 25, 2018.

[99] For a discussion of the centrality of NATO in U.S. strategic thinking, see Michael J. Mazarr, Miranda Priebe, Andrew Radin, and Astrid Stuth Cevallos, *Understanding the Current International Order*, Santa Monica, Calif.: RAND Corporation, RR-1598-OSD, 2016.

U.S. alliance commitments more contingent on allied burden-sharing.[100] Allies might therefore be concerned that failure to contribute to a mission that directly threatened the territory of a NATO member—a core function of NATO—would lead the United States to further question the value of the alliance.

Punishment by Noncoalition Allies

A military response in the Baltics could be conducted by a coalition of the willing made up of NATO and non-NATO partners acting outside of the NATO alliance. Although there is no legal restriction on independent action outside of NATO, some members see NATO's norms of consultation and consensus building as essential features of the alliance. As a result, they might seek to deter or punish those who would break that norm. Punishment by noncoalition allies seems unlikely in the event of a conventional attack on a NATO member. However, it could be more plausible in an unconventional scenario in which Russian intentions and Article 5 commitments are less clear.

Punishment by noncoalition allies seemed possible in the run-up to the Iraq War, when disputes over whether to support the U.S. drive to war with a "coalition of the willing" grew heated within Europe. For example, when the "Vilnius 10"—a group of Eastern European aspirants to the European Union (EU)—wrote a letter supporting U.S. operations, France threatened that such irresponsible behavior could cost them EU membership.[101]

Net Effect of Countervailing Pressures on An Ally's Decisionmaking

The range of factors driving allies' decisionmaking create countervailing considerations. Overall, allies will be more likely to participate in NATO operations if the public supports a military response or (in the absence of public support) there are no upcoming elections and elites agree on participation. A state is also more likely to participate in military action against Russia if it has a centralized foreign policy decisionmaking structure.

Perceptions of the threat posed by Russia and the risk of Russian retaliation would also influence each state's decisionmaking. Allies that view Russia as ambitious, opportunistic, or a threat to their homelands are more likely to participate. Concerns about military or economic retaliation might weigh more heavily on states that are less concerned about the threat Russia poses.

Finally, states will consider the value that they place on the alliance—and their role in it—as they deliberate over whether to engage in or support a military response.

---

[100] David E. Sanger and Maggie Haberman, "Donald Trump Sets Conditions for Defending NATO Allies Against Attack," *New York Times*, July 20, 2016.

[101] Gordon and Shapiro, 2004, pp. 133–134.

When a nation's goals for the operation are aligned with those of its allies and when it has a say in coalition decisions, a state will be more likely to participate. The value that the state places on the existence of the alliance—either for security or nonsecurity reasons—will help to determine its response to a given crisis.

The weight of each of these factors is likely to vary by individual state and by the circumstances surrounding a given scenario. In this chapter, we have offered a framework for planners to systematically assess each ally's willingness to contribute to a military response based on the specific context they are facing. The next two chapters illustrate additional insights that can be gained by systematically considering how these drivers of allies' decisionmaking might be influenced by the nature of a conflict—conventional or unconventional—and by specific activities undertaken by the Russian government.

CHAPTER THREE

Decisions to Contribute to a Military Response in Unconventional and Conventional Scenarios

We have identified three major categories of factors that could influence allied participation in a military response to Russian aggression: domestic politics, alliance dynamics, and perception of the Russian threat. As discussed in the previous chapter, the importance of these factors will vary depending on the context. We now consider how the relative importance of these factors could vary across two scenarios: an unconventional and a conventional Russian attack on a Baltic country.[1]

We do not focus on the details of one particular Baltic scenario. Instead, we identify key features of potential Russian unconventional and conventional strategies that could shape Baltic conflict scenarios. The key variables in these scenarios that are likely to shape NATO allies' decisions on whether to respond are the risk of escalation and potential for Russian retaliation. Both factors are compounded by the level of allied uncertainty about Russian government involvement and intentions and the applicability of Article 5.

Overall, the analysis described in this chapter suggests that domestic political factors are likely to weigh more heavily in an unconventional scenario. Uncertainty about the nature of unfolding events could make assessments about the threat Russia poses and alliance dynamics more complicated. As a result, elites might disagree, making it hard to build a domestic consensus on a military response. In a conventional scenario, heightened perceptions of a threat from Russia and alliance considerations are more likely to overwhelm these domestic factors.[2]

---

[1] Russian aggression in the Baltics, either conventional or unconventional, are the most commonly discussed scenarios for a NATO confrontation with Russia. See Loren B. Thompson, "Why the Baltic States Are Where Nuclear War Is Most Likely to Begin," *National Interest*, July 20, 2016; David A. Shlapak and Michael Johnson, *Reinforcing Deterrence on NATO's Eastern Flank: Wargaming the Defense of the Baltics*, Santa Monica, Calif.: RAND Corporation, RR-1253-A, 2016.

[2] Domestic political factors tend to matter most during periods of uncertainty, when threat perceptions are lower, and when a state is dealing with peripheral, rather than vital, national security interests (Barry Posen, *The Sources of Military Doctrine: France, Britain, and Germany Between the World Wars*, Ithaca, N.Y.: Cornell University Press, 1984; Snyder, 1984, p. 30).

Unconventional Scenario

Analysts have suggested that an unconventional scenario could begin with Russian support to antigovernment groups in a Baltic state. For example, one analyst suggested that Russia could encourage Russian populations to "begin rioting, protesting for their rights, claiming to be persecuted, asking for 'international protection.' A suspiciously well-armed and well-trained 'Popular Front for the Liberation of the Russian Baltics' would then appear."[3] This force could be composed of nonuniformed Russian unconventional forces—the infamous "little green men" used in Crimea and eastern Ukraine.[4] Survey research has indicated that ethnic Russian populations in the Baltic states are significantly better integrated and committed to their European identity than Ukraine's ethnic Russians.[5] However, given the potential for alienation within the Russian-speaking community, this could be one plausible element of a future scenario. Whatever the specifics, allies' decisionmaking is likely to be affected by the unique features of an unconventional scenario.

Perceptions of a Russian Threat

During an unconventional scenario, the Russian government's responsibility for an attack might be ambiguous. In recent operations, including in Eastern Ukraine, Russia has used proxies, clandestine operations, and even cyberattacks by nongovernment sources to obscure Russian government involvement.[6] Moreover, as it has done in the recent past, Russia would likely pursue aggressive information operations aimed at European leaders and publics to sow further doubt about Russian involvement.[7]

During an unconventional scenario, it might be difficult to obtain accurate information about Russian activities or, if only highly classified intelligence is available, to share it widely. Ambiguity about Russia's role in an unconventional scenario means that foreign policy elites might make divergent assessments about the threat Russia poses and its importance relative to other national security concerns. Since information

---

[3] Paul D. Miller, "How World War III Could Begin in Latvia," *Foreign Policy*, November 16, 2016.

[4] Andrew Stuttaford, "Watching the Baltic—'Little Green Men' and Other Concerns," *National Review*, February 8, 2015.

[5] Jill Dougherty and Riina Kaljurand, *Estonia's "Virtual Russian World": The Influence of Russian Media on Estonia's Russian Speakers*, Tallinn, Estonia: International Centre for Defence and Security, October 2015.

[6] For discussions of Russia's use of these tools in the past, see Ron Thornton and Manos Karagiannis, "The Russian Threat to the Baltic States: The Problems of Shaping Local Defense Mechanisms," *Journal of Slavic Military Studies*, Vol. 29, No. 3, 2016; Miller, 2016; Fletcher School, Tufts University, "SIMULEX 2016: Crisis in the Baltic Region and the Middle East," crisis management exercise, October 21–22, 2016; Stephen Blank, "Putin's Next Potential Target: The Baltic States," *RealClear Defense*, January 5, 2016. For challenges associated with conclusively attributing cyberattacks, see Thomas Rid and Ben Buchanan, "Attributing Cyber Attacks," *Journal of Strategic Studies*, Vol. 38, No. 1–2, December 2015.

[7] Clark et al., 2016.

during an unfolding unconventional scenario might be difficult to interpret, precrisis perceptions of Russia might have the most significant impact. For example, states that perceive the scenario as part of a broader Russian militarism that should be immediately stemmed—e.g., Poland—might be more ready to act.[8] Conversely, states that worry that NATO actions might unnecessarily provoke a Russian response, such as Bulgaria, might be more hesitant.[9]

Alliance Politics

Ambiguity about Russia's role and a lower level of violence could lead to intense disagreement within the alliance about the applicability of Article 5. If several thousand uniformed Russian soldiers invaded Baltic territory, the applicability of Article 5 would be clear. However, the alliance does not currently have a shared understanding of which unconventional attacks meet the standard for Article 5 commitments. In its most definitive statement, NATO heads of state announced at the Wales Summit that a cyberattack could be considered for Article 5 on a "case-by-case basis."[10] Ambiguity surrounding alliance commitments and Russian involvement make it less likely that coalition members would punish nonparticipants. Alliance considerations therefore are likely to be less prominent in this scenario than in a conventional scenario.

Domestic Politics

For allies with a low level of concern about Russia prior to a crisis, domestic politics might weigh very heavily in decisions about whether to respond to reports of Russian unconventional operations. With incomplete information, domestic elites might have varying interpretations of events, making it unlikely that they will agree that the Russian threat is unambiguously growing. Even if leaders agree on the threat, it might be difficult to build a clear case that can sway a skeptical public. In the months and weeks prior to a national election, public opposition might weigh particularly heavily. Unlike a conventional attack, which would present a clear threat to the alliance and perhaps even the nation, the ambiguities inherent in unconventional attacks might not be compelling enough for national leaders to look beyond potential electoral implications.

---

[8] See Piotr Buras and Adam Balcer, "An Unpredictable Russia: The Impact on Poland," European Council on Foreign Relations, July 15, 2016.

[9] "Bulgaria Says Will Not Join Any NATO Black Sea Fleet After Russian Warning," 2016.

[10] NATO, "Wales Summit Declaration," transcript, September 5, 2014b. Analysts have since noted the extent to which the declaration of Article 5 in response to a cyber incident would be situationally dependent, leaving a gap in NATO doctrine. See Jarno Limnéll and Charly Salonius-Pasternak, *Challenge for NATO—Cyber Article 5*, Stockholm: Center for Asymmetric Threat Studies, Swedish Defense University, briefing paper, June 2016.

Conventional Scenario

A conventional conflict with Russia could result from escalation of an unconventional scenario or could begin without an unconventional phase. Some have argued that Russia could stage a large exercise and then divert some forces to make a rapid land grab of a small part of a Baltic state.[11] More-ambitious scenarios include Russia seizing a capital city and large parts of one of the Baltic states.[12] Russia would likely continue to use information operations in a conventional scenario to justify Russian behavior or threaten retaliation.[13] Unlike an unconventional scenario, a conventional scenario would be marked by far greater clarity about Russia's role and the applicability of Article 5.

Perceptions of a Russian Threat

A conventional scenario would include fewer questions about Russian involvement because of observable realities on the ground. There would still be some questions about Russia's intent and aims, especially if Russia conducted an information campaign justifying its actions. Still, decisionmakers would have greater clarity about the threat than in an unconventional scenario.

At the same time, a military response in a conventional scenario would carry greater risks of escalation and Russian retaliation. Responding to a conventional attack could involve air and ground attacks on Russian forces or attacks on air defense and other military capabilities in the Russian territory of Kaliningrad. Russia might take such steps as aggressive bomber runs into allied air space to demonstrate Russian capability and willingness to impose costs upon any allies participating or indirectly supporting NATO defensive efforts.[14] With direct clashes between Russian and allied forces and greater military activity in general, the potential for escalation would be high.[15] Allies would face a stark choice between a greater risk of escalation and Russian retaliation and a desire to respond to Russian aggression.

---

[11] Keir Giles, *Russia's "New" Tools for Confronting the West: Continuity and Innovation in Moscow's Exercise of Power*, London: Chatham House, the Royal Institute of International Affairs, March, 2016. Others test more maximalist assumptions, including the RAND Corporation study based on conventional wargames that highlight Russian capacity to march to Baltic capitals within in 60 hours (Shlapak and Johnson, 2016).

[12] Shlapak and Johnson, 2016.

[13] Potential justifications for Russian military activity on Baltic soil range from ensuring access to the Russian enclave of Kaliningrad to protecting vulnerable ethnic minorities (Edward Lucas, *The Coming Storm: Baltic Sea Security Report*, Washington, D.C.: Center for European Policy Analysis, June, 2015, p. 15; Clark et al., 2016; Richard Shirreff, *War with Russia: An Urgent Warning from Senior Military Command*, New York and London: Quercus, 2016). See also Jeffrey Rathke, "Can NATO Deter Russia in View of the Conventional Military Imbalance in the East?" Center for Strategic and International Studies, November 30, 2015.

[14] See Damien Sharkov, "Russian Supersonic Jets Force NATO to Scramble Fighter Jets," *Newsweek*, March 25, 2015.

[15] Shlapak and Johnson, 2016.

Alliance Politics

Because a conventional attack would be attributable to Russia and would constitute an "armed attack," it would clearly qualify for Article 5 response. Should the target state announce that it was experiencing an Article 5 situation and request NATO assistance, NATO's failure to respond would be seen by many, if not all, members as a failure of the alliance's core function.[16] In most cases, allies would view a decision to respond to a conventional Russian attack as an existential question for the alliance.

For frontline states that rely on NATO for their security, alliance considerations would outweigh all other factors. Senior officials at the Lithuanian Foreign Ministry have stated publicly that "Vilnius is indeed ready to defend Warsaw. In the opposite case, NATO would lose its credibility."[17] Similarly, a Polish opinion poll found that 71 percent of Poles felt that their country should defend the Baltic states if they were attacked.[18] While these states would contribute, they would be limited by a need to hold back forces to defend their own territories and populations in case Russia retaliated or turned in their direction next.[19]

Even states facing less acute threats to national security would feel strong alliance pressures to contribute. These states might strongly value NATO for security reasons and worry it would dissolve in the face of this type of challenge. Moreover, although there has been minimal alliance punishment for failure to participate in out-of-area operations, NATO members might respond more harshly for failure to uphold mutual defense commitments within Europe.

Though there would be strong pressures to join, and these would overcome domestic opposition under many circumstances, there is a countervailing consideration that would be even more important in the conventional scenario. Some allies might fear that maximalist goals by the United States or other allies could embroil them in a wider conventional war—or even in a nuclear war—with Russia. Certainly, the prospect of U.S.-driven regime change has been a concern within the Russian government, and Moscow could cite it as a covert U.S. objective.[20] To the extent that the United States would be willing to commit itself to collective decisionmaking structures, this concern could be mitigated. This concern might be weaker in the context of a NATO operation in which procedures and decisionmaking processes are already established than concern would be in a coalition of the willing.

---

[16] Although Article 5 technically applies regardless of the cause of the attack, clear evidence of NATO members' culpability in provoking a Russian attack might complicate discussions about the applicability of Article 5.

[17] Rolandas Kacinskas, political director at the Lithuanian Foreign Ministry, quoted in "Lithuanian Diplomats Tell Poland: Vilnius Ready to Defend Warsaw," *LETA*, September 27, 2016.

[18] "Some 71% of Poles Think Poland Should Defend Baltics in Case of Attack," *Delfi by Lithuania Tribune*, November 10, 2016.

[19] Shlapak and Johnson, 2016.

[20] Defense Intelligence Agency, 2017.

Clarity of decisionmaking would be complicated by allied complicity—perceived or actual—in provoking the conflict. For this reason, Russian information operations during the opening phases of a conventional conflict would likely focus heavily on the role of the attacked ally in triggering the crisis, as well as the legitimation of Russian actions. As will be discussed in Chapter Four, Russian information operations actively sought to characterize Georgia and Ukraine as complicit during Russian military interventions.

Domestic Politics

The clarity of a conventional scenario could make domestic opposition less constraining on decisionmakers than it would be in the unconventional scenario. First, the public might be more supportive of a military response to a blatant Russian attack. Second, the government would have greater ability to construct a narrative to build support for a military response and potentially undermine Russian disinformation campaigns. Third, government elites might be more likely to come to a common view of the threat that Russia poses and less likely to use foreign policy to win electoral points. To the extent that domestic politics is important in this scenario, it is most likely to affect the extent or nature of a state's contribution.

Conclusion

The factors outlined in Chapter Two are likely to affect allies' decisions about contributing to collective defense in any context. However, the weight that allies place on these factors will depend on the scenario. In an unconventional scenario, uncertainty about Russia's role and the applicability of Article 5 might bring domestic political considerations to the foreground. Public opposition, especially for an electorally vulnerable government, will make a military contribution more contentious. In a conventional scenario, domestic politics probably will matter less, as alliance dynamics and greater clarity about Russia's intentions become more-powerful factors in allies' decisions.

CHAPTER FOUR

Possible Modes of Russian Influence on Allies' Decisionmaking

Most of the factors affecting decisions to participate in an allied military response are not static. Domestic political changes can affect some factors, such as government electoral vulnerability on foreign policy, and evolving alliance politics can alter perceptions of the risk of entrapment. Some factors are also vulnerable to adversary manipulation. The Soviets practiced "reflexive control," which aimed to disrupt decisionmaking through disinformation and deception; Russia has employed this tactic more recently during conflicts and foreign policy clashes.[1] In this chapter, we use the framework developed in Chapter Two to identify ways that Russia could manipulate NATO member decisions about military responses. We identify strategies that Russia could employ in peacetime, as well as steps that Russia could take at the outset of an unconventional or conventional conflict with NATO. While some of these strategies are hypothetical, we also provide recent examples of actual Russian influence attempts.

Possible Modes of Russian Influence on Domestic Politics

There are a number of mechanisms by which Russia might seek to influence the domestic politics in NATO countries and, in turn, their decisionmaking on military responses (see Table 4.1). These mechanisms include shaping public attitudes and engaging with and funding sympathetic politicians in NATO countries. Russia has reportedly used all three of these tools in recent years.

Influencing Public Attitudes

Russia has already sought to influence national decisionmaking by targeting public opinion in NATO countries.[2] Russia appears to have taken two approaches in its targeting of public attitudes. The first is to present the public with Russian perspectives on

---

1 Marta Kepe, "NATO: Prepared for Countering Disinformation Operations in the Baltic States?" *RAND Blog*, June 7, 2017.

2 NATO's Strategic Communications Centre of Excellence has documented Russia's doctrinal and organizational changes to more effectively connect with Western publics since the 2008 Georgian war.

Russia has also sought to affect foreign policy choices of NATO members by providing financial and other assistance to far-right European parties that are skeptical of European institutions and activist foreign policies.[7] In the run-up to Montenegro's 2017 accession to NATO, Russian government media outlets in Montenegro unsuccessfully backed an anti-NATO Montenegrin political coalition; anti-NATO movement leaders were reportedly on payroll lists of Russian institutions in Serbia, such as the Russian media outlet *Sputnik*.[8] Russia could also make in-kind contributions, such as launching a social media campaign against a pro-Russian candidate's opponents.[9] By supporting these parties, Russia's goal is to get politicians elected who are less likely to adopt a military response toward Russia. Even if these candidates do not win elections, Russian support could give them the opportunity to publicize foreign affairs in general and Russia's positions specifically, increasing the impact of public opposition.

Possible Modes of Russian Influence on Allies' Perceptions

Russia could try to manipulate NATO members perceptions in two ways. First, Russia might wish to mitigate some NATO allies' view of Russia as a threat by denying or justifying its actions. Second, Russia might threaten retaliation against vulnerable countries to deter them from joining a military response. Attempting to both reassure and deter simultaneously can be difficult, so Russia would probably prioritize one approach over the other.

Reassurance About Russian Aims and Motivations

The Russian government has sought to assuage concerns about its intentions in past crises by denying expansive aims and obscuring its role by using proxies. When Ukraine's foreign ministry warned that "in Crimea, a general Russian passportization is gaining

---

[7] For example, Russia reportedly provided a loan of 9.4 million euros to Marine Le Pen's National Front party, which has taken a pro-Russian stance on foreign policy. Le Pen has been a vocal opponent of Western sanctions against Russia, stating that she "absolutely disagree(s) that (Crimea) was an illegal annexation: a referendum was held and residents of Crimea chose to rejoin Russia" (Tom Batchelor, "Marine Le Pen Insists Russian Annexation of Crimea Is Totally Legitimate," *The Independent*, January 3, 2017; Janis Sarts, "Russian Interference in European Elections," Hearing Before the Select Committee on Intelligence, U.S. Senate, Washington, D.C., June 28, 2017).

[8] Vesko Garčević, "Russian Interference in European Elections," hearing before the U.S. Senate Select Committee on Intelligence, Washington, D.C., June 28, 2017.

[9] Evidence is inconclusive, but some have argued that Russia provided such support to President Miloš Zeman of the Czech Republic (Marc Santora, "Czech Republic Re-Elects Milos Zeman, Populist Leader and Foe of Migrants," *New York Times*, January 27, 2018; Jakub Janda and Veronika Víchová, "The Kremlin's Hostile Influence in the Czech Republic: The State of Play," *Warsaw Institute Review*, Warsaw, Poland: Warsaw Institute, August 10, 2017; Jakub Janda, "How Czech President Miloš Zeman Became Putin's Man," *The Observer*, January 26, 2018; Keno Verseck, "Is the Czech Republic Moving Closer to China and Russia?" *Deutsche Welle*, January 31, 2018).

ground," the Kremlin declared its respect for the territorial integrity of Ukraine.[10] After seizing Crimea, Putin denied that Moscow had plans to annex Crimea but then went on to do so.[11]

Another approach has been to build ambiguity and plausible deniability by using proxies and nonuniformed Russian forces. In the days following the takeover of Crimea by Russian forces, Putin denied Russian involvement and claimed that the influx of armed forces in Crimea were exclusively "local self-defense forces."[12] The Russian government officially maintained this position for several weeks, until President Putin conceded in an April 2014 interview that Russian troops had indeed "stood behind Crimea's self-defense forces."[13] Throughout the subsequent Russian campaign in eastern Ukraine, Russia sought to minimize its perceived role in the conflict through the use of proxies, nonuniformed personnel, and other ambiguous tactics.[14] The approach contributed to divergent intelligence assessments across the alliance.[15]

In a conventional scenario, denials would not be effective. Instead, as it has done in the past, Russia might appeal to international norms to legitimize its actions. For example, the Russian decision to organize a referendum in Crimea appeared to be an attempt to legitimize the annexation with evidence of popular support. Similarly, Russia has sought to use peacekeeping mandates to maintain an estimated 2,000-strong force in the Moldovan breakaway region of Transnistria.[16]

Threats of Military Retaliation and Escalation

Russian government officials have previously issued public threats against foreign states that have acted in opposition to Russian interests and would likely do so again in the event of a crisis with NATO. At times, such threats have been vague: After Montenegro's decision to join the NATO alliance, Foreign Minister Sergey Lavrov warned cryptically that "in response to the hostile policy chosen by the Montenegrin authorities, the Russian side reserves the right to take retaliatory measures on a reciprocal

---

[10] Paul A. Goble, "Russian 'Passportization,'" *New York Times*, September 9, 2008.

[11] Bill Chappell and Mark Memmott, "Putin Says Those Aren't Russian Forces in Crimea," *NPR*, March 4, 2014.

[12] Michael Kofman, Katya Migacheva, Brian Nichiporuk, Andrew Radin, Olesya Tkacheva, and Jenny Oberholtzer, *Lessons from Russia's Operations in Crimea and Eastern Ukraine*, Santa Monica, Calif.: RAND Corporation, RR-1498-A, 2017. Putin claimed instead that the armed men in Crimea represented "local self-defense forces" (Chappell and Memmott, 2014).

[13] Alexei Anischchuk, "Putin Admits Russian Forces Were Deployed to Crimea," *Reuters*, April 17, 2014.

[14] Karoun Demirjian and Michael Birnbaum, "Russia Escalates Tensions with Aid Convoy, Reported Firing of Artillery Inside Ukraine," *Washington Post*, July 22, 2017.

[15] "Breedlove's Bellicosity: Berlin Alarmed by Aggressive NATO Stance on Ukraine," *Spiegel Online*, March 6, 2015.

[16] Eugen Tomiuc and Radu Benea, "Russia Objects to Moldovan Call for Removing Troops from Transdniester," *Radio Free Europe/Radio Liberty*, August 23, 2017.

basis. In politics, just as in physics, for every action there is an opposite reaction."[17] At other times, threats have been more direct. In 2016, Putin publicly warned that Poland and Romania would be in Russia's "crosshairs" because of their decision to host U.S. missile defense elements, emphasizing Russian use of sea-based missiles in Syria and Russia's Iskander missiles.[18]

Russian nuclear capabilities would loom large over any conventional conflict between NATO and Russia, and Russian signaling of its willingness to use these weapons would raise concerns about potential nuclear escalation. Since the beginning of the Ukraine crisis, President Putin has twice warned NATO members not to provoke a nuclear-armed Russia, stated that he was ready to put nuclear weapons on alert, and held recent exercises incorporating nuclear escalation.[19] In response to Russian nuclear posturing in 2016, Germany's then–Foreign Minister Frank-Walter Steinmeier noted that he was watching the collapse of relations between the West and Russia with dismay, and warned that tensions between the two were "more dangerous" at that point than during the Cold War.[20]

Exploiting Economic Dependence

In the past, Russia has sought to increase other states' dependence on Russia, then later threaten sanctions to influence their decisionmaking. Through below-market pricing and infrastructure development, Russia has become the primary supplier of natural gas to Eurasian and eastern European states.[21] New pipelines to Europe will mean that Russia can bypass these states to maintain supply for lucrative sales to Western Europe—even when gas has been cut off to a Eurasian or eastern European state.[22] As a result, Russia has fostered dependency on Russian gas without making Russia dependent on these states in return. In 2014, Russia used this leverage to coerce changes in Ukraine's foreign policy orientation.[23] When Ukraine's neighbors sought to mitigate the impact by providing "reverse flow" from their own supplies, Putin threatened to—and later did—reduce supplies to those who participated.[24] For example, Russia

---

[17] Dusica Tomovic, "Russia Threatens Retaliation After Montenegro Joins NATO," *BalkanInsight*, June 6, 2017.

[18] Dyomkin, 2016.

[19] Kathrin Hille, "Russia Attacks NATO's Decision to Strengthen Defences in Poland," *Financial Times*, July 10, 2016.

[20] Solovyov and Systas, 2016.

[21] Ryan C. Maness and Brandon Valeriano, *Russia's Coercive Diplomacy: Energy, Cyber, and Maritime Policy as New Sources of Power*, Basingstoke, U.K.: Palgrave Macmillan UK, 2015, p. 118.

[22] Examples include the failed Nordstream and South stream initiatives, and the current Nordstream II initiative, which will employ an expensive underwater pipeline to permit Russia to provide gas directly to Germany, bypassing Ukraine and the Baltic states (Maness and Valeriano, 2015, p. 118).

[23] "U.S.: Russia Uses Energy Supplies 'to Control Ukraine,'" 2014.

[24] Vladimir Putin, "Answers to Journalists' Questions," *Kremlin.ru*, June 6, 2014.

reduced Polish gas supplies by 45 percent in fall 2014, forcing the Polish government to drop the policy of providing reverse flows to Ukraine.[25] In another example, Russia applied agricultural and tourism sanctions on Turkey to force a Turkish apology after a Russian fighter was shot down in Turkish airspace. Sanctions might be part of the reason that President Recep Tayyip Erdogan apologized for the incident seven months later.[26]

In other cases, Russia's use of economic coercion has had mixed results. For example, after the EU imposed sanctions on Russia for Russian activities in eastern Ukraine, Russia responded with countersanctions. Italy, which saw its transport equipment exports drop by 42 percent after the imposition of sanctions, has continued to participate in the sanctions regime but has sought to limit its scope.[27] Although Iceland's government and primary political opposition parties have remained committed to sanctions, Russian countersanctions on Iceland's fishing industry have reportedly prompted some internal questions as to the wisdom of maintaining the policy.[28] In spite of the costs to EU members, as of September 2018, the EU has continued to extend sanctions against Russia every six months. Notably, those countries hardest hit by sanctions—the Baltic states and Poland—are among the most-ardent supporters of sanctions.[29]

Over time, Russia could seek to take more-aggressive steps to increase the financial and economic dependence of NATO members, making them more vulnerable to threats of economic retaliation during a crisis. However, given that Russia has had mixed results with sanctions in the past, threats to NATO members' energy, trade, tourism, and financial relationships with Russia would not likely be the primary drivers of ally decisions about a military response.

---

[25] Marek Strzelecki and Maciej Martewicz, "Gazprom Limits Polish Gas Supplies as Reverse Flows Halt," *Bloomberg*, September 10, 2014.

[26] Jim Zanotti and Clayton Thomas, *Turkey: Background and U.S. Relations*, Washington, D.C.: Congressional Research Service, R41368, August 31, 2018, pp. 35–36; "Russia, Turkey Sign Gas Pipeline Deal," *Radio Free Europe/Radio Liberty*, October 10, 2016. Turkish news reports cited loss of tourism revenues ranging from $5 billion to $10 billion (Asli Aydıntaşbaş, *With Friends Like These: Turkey, Russia, and the End of an Unlikely Alliance*, London: European Council on Foreign Relations, June 23, 2016).

[27] Marcin Szczepański, *Economic Impact on the EU of Sanctions over Ukraine Conflict*, Brussels: European Parliamentary Research Service, PE 569.020, October 2015; Andrew Rettman, Peter Teffer, Eric Mauice, and Eszter Zalan, "Italy Shields Russia from EU Sanctions Threat," *EUobserver*, October 21, 2016.

[28] Lowana Veal, "Iceland: Fish Fight Spawned over Ukraine Sanctions," *Al Jazeera*, October 8, 2015; "Russian Embargo: Reactions," *Iceland Monitor*, August 14, 2015.

[29] David M. Herszenhorn, "Putin Extends Counter-Sanctions Against EU," *Politico*, June 30, 2017.

Possible Modes of Russian Influence on Alliance Politics

Russia could also seek to manipulate how allies perceive one another. Over the longer term, Russia might try to erode the perceived value of the NATO alliance in the eyes of member state publics and elites.

Blame Target State for Real or Fabricated Provocations

In the lead-up to and during a crisis, Russia would highlight provocations by NATO members by conducting information operations that might or might not be accurate.[30] Russia has used this approach in the past. For example, in 2008, Russia mounted a major information campaign to label Georgian authorities as the instigators of the crisis between the two countries.[31] Moreover, Russian television broadcast images of U.S. equipment from the U.S.-Georgian bilateral exercise Immediate Response 2008 as "proof" that U.S. military training and equipment lay behind the Georgian intervention in South Ossetia.[32] Similarly, throughout Russian military operations in Ukraine, Russian information campaigns pointed to shelling of Donbass civilians by the Ukrainian government, purported the genocidal ambitions of the Ukrainian government, denied that Ukraine is an independent state ruled by a legitimate government, and even claimed the radical Islamization of Ukraine.[33] Russian social media themes and statements by Russian leaders also blamed expansive U.S. aims in Europe for provoking the Ukraine conflict.[34] Efforts to highlight differences in allied goals could be entirely fabricated. For example, a 2018 RAND study identified pro-Russia websites in the Czech Republic that pushed articles purporting NATO plans to attack Russia from eastern Europe in the absence of approval from local governments.[35]

In a Baltic scenario, Russia might highlight Baltic state repression of ethnic Russians to justify its intervention. If allies accept this narrative, they might be less likely to contribute to collective defense, either because they see the target country as culpable or because of concerns that the target state will engage in future provocations that make escalation more likely.

---

[30] Kepe, 2017.

[31] Ariel Cohen and Robert E. Hamilton, *The Russian Military and the Georgia War: Lessons and Implications*, Carlisle Barracks, Pa.: Strategic Studies Institute, U.S. Army War College, June 2011.

[32] Cohen and Hamilton, 2011.

[33] Denis Bohush and Oleksandra Baglai, *Russian Propaganda Concerning Ukraine During the Syrian Campaign: An Innovative Approach to Assess Information Activities*, Riga, Latvia: NATO Strategic Communications Centre of Excellence, November 2016.

[34] Kofman et al., 2017. For an example of Russian leader statements blaming the United States, see "Lavrov Blames NATO for Russian Aggression in Ukraine," *UNIAN*, February 19, 2018.

[35] Helmus et al., 2018, p. 22.

Claims of Divergent Alliance Goals

Russia could also seek to promote a perception that allies do not share the same goals more generally through peacetime information operations. Recent Russian disinformation efforts in Sweden show that such activities are already underway. A 2018 Brookings study on the future of political warfare identified a tsunami of fake stories in Swedish news about the downsides of entering NATO,

> including untruthful claims about the alliance plotting to stockpile nuclear weapons on Swedish soil, NATO's prerogative to attack Russia from Swedish territory without Stockholm's consent, and NATO soldiers having license to sexually assault Swedish women without fear of prosecution because of legal immunity.

These stories originated from Russian sources.[36]

Russian efforts to undermine NATO as an institution are well documented. NATO international staff have catalogued 32 common Russian myths about the alliance, including purported NATO aggression toward Russia and claims regarding failures of specific NATO operations and activities.[37]

Many of Russia's options for manipulating allies' decisionmaking about collective defense rely on influence operations that seek to alter the perceptions and opinions of NATO publics and elites. We have highlighted examples of core themes that allies might anticipate that Russia would promote to encourage public opposition, sow doubts about alliance unity, and deter military contributions. The next chapter discusses ways that the United States and its allies can counter these influence attempts.

---

[36] Alina Polyakova and Spencer B. Boyer, *The Future of Political Warfare: Russia, the West, and the Coming Age of Global Digital Competition*, Washington, D.C.: Brookings Institution, March, 2018; Neil MacFarquhar, "A Powerful Russian Weapon: The Spread of False Stories," *New York Times*, August 28, 2016.

[37] Foo Yun Chee, "NATO Says It Sees Sharp Rise in Russian Disinformation Since Crimea Seizure," *Reuters*, February 11, 2017. For the NATO website, please see NATO, "NATO-Russia Relations: The Facts," webpage, September 7, 2018d.

CHAPTER FIVE

Conclusion and Recommendations to Promote Contributions to Collective Defense

In this chapter, we use the decisionmaking framework developed in Chapter Two to identify policies that could ameliorate allies' concerns and increase the likelihood that allies will support a military response in the event of a Russian attack. Earlier chapters highlighted reasons that allies might be reluctant to contribute to collective defense. Some are based on domestic political and alliance dynamics; others might result from Russian manipulation. The recommendations provided in this chapter include steps that the United States should take in peacetime or to build support for a U.S.-led military response during a crisis. Table 5.1 summarizes the recommendations that are detailed in this chapter.

U.S. policy cannot easily affect all of the domestic drivers of allies' decisionmaking. Foreign policy decisionmaking structures are largely fixed. However, efficiently releasing U.S. intelligence on Russian activities to the public could help to build elite consensus to counter Russian attempts to manipulate public opinion. Proposing additional ways for NATO to engage with opposition parties and emerging leaders could also have limited effects on elite consensus about military responses in the longer term.

Provide Accurate and Timely Information on Russian Activities to NATO Publics

U.S. policy officials are aware of Russian attempts to shape public opinion and have periodically provided declassified intelligence analysis on Russian activities to the public, including on Russia's role in the 2014 shootdown of a Malaysian Airlines commercial aircraft.[1] In another example, in 2017, the Defense Intelligence Agency published an unclassified assessment of Russian military capabilities, strategy, and tactics.[2]

One way to expand public outreach would be to provide additional financial support to both government and nongovernment entities designed to facilitate access to critical information to public audiences across the alliance and beyond. During the

---

[1] See Elizabeth Chuck, "White House to Lay out U.S. Intelligence on MH17," *NBC News*, July 22, 2014.

[2] Defense Intelligence Agency, 2017.

Table 5.1
U.S. Policies to Reduce Allies' Vulnerability to Manipulation and Promote Allies' Contributions to Collective Defense Operations

Category	Driver of Allies' Decisionmaking	Possible U.S. Policies
Domestic politics	• Public opinion about military responses and alliance commitments	• Provide accurate and timely information on Russian activities to NATO publics
	• Vulnerability of governing coalition on foreign policy issues	• Expand outreach to political elites on NATO's enduring value—e.g., through NATO's Parliamentary Assemblies and other political exchange programs
	• Foreign policy decision-making structure	• Not easily influenced by U.S. policy
Perceptions of Russian threat	• Perception of Russian aims and motivations	• Develop and release more U.S. intelligence on Russian activities to NATO allies • Provide more U.S. personnel to NATO's intelligence and analysis organizations • Initiate dedicated dialogue with allies on Russian aims and motivations
	• Competing national security demands	• Not easily influenced by U.S. policy
	• Vulnerability to Russian military retaliation and concern about risks of escalation	• Encourage allies to enhance civil defense preparations, including from cyberattack • Encourage allies to invest in point and small-area missile defenses
	• Vulnerability to Russian economic retaliation	• Propose NATO dialogue with the EU on diversifying European energy supplies • Consider developing a joint NATO-EU strategic reserves program
Alliance politics	• Participation of other allies	• Indirectly affected by other policies that increase the number of coalition participants
	• Alignment of goals among participants	• Encourage Baltic states to continue to address their Russian populations' grievances to reduce allies' concerns about entrapment • Provoke discussion of political goals by using conventional and unconventional Russia scenarios in crisis-management exercises
	• Ability to restrain coalition members	• Sustain high-level NATO political exercises to build confidence in decisionmaking processes
	• Consequences of alliance dissolution or abandonment	• Consistently signal U.S. commitment to the alliance so allies are confident that they will continue to get benefits of collective defense if they participate • Communicate expectations of participation and clear consequences for failing to support an ally in the face of an attack
	• Punishment by noncoalition allies	• Indirectly affected by other policy changes that reduce the number of nonparticipants

Cold War, Radio Free Europe/Radio Liberty and Voice of America played this role; the United States could consider reinvesting in these and other public outreach channels.[3]

U.S. policymakers could also make better use of publicly available commercial sources. In April 2014, NATO disseminated photographs taken by the commercial satellite imaging company DigitalGlobe, providing some of the first public evidence of U.S. claims of a large-scale Russian conventional military buildup on the Ukrainian border.[4] During future crises, U.S. and NATO policymakers should continue to identify and point to open-source data, such as that provided by DigitalGlobe, as one way to circumvent challenges that NATO and the United States have in sharing and disseminating classified information. The hybrid branch of the NATO Joint Intelligence and Security Division (JISD) and the Hybrid Center of Excellence, both established in 2017, could provide opportunities for the analysis and rapid dissemination of credible open-source and nongovernmental intelligence products.

Finally, the United States could devote additional resources to more rapidly sanitize and declassify more intelligence for public release. As discussed in Chapter Three, Russia can more easily deny involvement in unconventional attacks, and public opinion might have more influence on allies' decisions in these scenarios. Therefore, even though additional personnel and training would be needed to revamp public release processes, these resources might be particularly valuable if an unconventional Russian attack seems likely. A dedicated Intelligence Community task force focused on identifying and sanitizing intelligence relating to Russian activities for public release—as well as for allied governments—could help to ensure that more-relevant information would be shared.

Invest in Additional Mechanisms for Engaging Opposition Parties and Future Political Leaders

When a governing party or coalition is electorally vulnerable, it is more likely to be sensitive to public opposition and concerned about elite consensus on foreign policy. NATO peacetime activities to engage with a range of political leaders in member countries, such as through the NATO Parliamentary Assembly, might lay the groundwork for elite consensus in a crisis. U.S. political leaders, particularly the U.S. ambassador to NATO, could increase U.S. involvement in these fora and emphasize their importance through both diplomatic and public engagement. The United States could also propose new NATO outreach initiatives. For example, NATO has centers of excellence in many member countries. Inviting parliamentary delegations and rising lead-

---

[3] The Russian government named Radio Free Europe/Radio Liberty and Voice of America as foreign agents in 2017; a Moscow court fined the organizations in July 2018 for failing to comply with the law. The organizations publicly dispute the characterization of their work ("U.S.-Funded Radio Free Europe/Radio Liberty Fined by Russia," *U.S. News & World Report*, July 5, 2018).

[4] David M. Herszenhorn, "Satellites Show Russian Forces Poised Near Ukraine," *New York Times*, April 10, 2014.

ers to conferences at these centers would be one way to engage with them on perceptions of Russia and to extend outreach beyond the usual visits to NATO headquarters and Supreme Headquarters Allied Powers Europe.[5] Recruiting participants beyond the transatlantic consensus could be challenging: In recent years, nationalist and populist political figures that question the value of NATO have gained greater prominence, and these politicians might not be willing to participate in NATO engagement initiatives. Without their participation, these activities might have more-limited effects than other U.S. policy options.

Perceptions of Russian Threat and Vulnerability to Retaliation

Increase Intelligence-Sharing on Russian Activities

When allies share a common view of the extent and range of Russian malign activities, they might be more likely to agree on the threat that Russia poses. Reaching a shared view of the threat is likely to be particularly challenging in an unconventional scenario, in which Russia's involvement might not be overt. Improving intelligence-sharing and fusion could prove critical in reducing ambiguity about the threat Russia poses and helping allies agree on a common approach both before and during a crisis. This section outlines some specific steps that the United States could take to promote a shared understanding of Russian activities.

Develop and Release More U.S. Intelligence on Russian Activities to NATO Allies

The United States already shares a significant number of intelligence products with NATO on security topics of interest to the allies. However, additional resources could be devoted to the development of intelligence analysis on Russian activities (including military, political, unconventional) by the U.S. Intelligence Community with the explicit purpose of releasing it to NATO allies.

Establishing a process to ensure that U.S. intelligence agencies can rapidly review relevant intelligence products and put it in a form that is releasable to NATO allies could help allies come to agreement about the threat. The United States already shares intelligence with NATO allies through NATO-wide agreements, bilateral agreements, and narrower multilateral arrangements (such as the "five eyes" [FVEY] agreement among the United States, United Kingdom, Canada, Australia, and New Zealand).[6] While the

---

[5] NATO members currently run about two dozen nationally based NATO centers of excellence, which could be leveraged to convene political leaders, parliamentary delegations, midlevel government officials, and rising leaders from both sides of the Atlantic.

[6] National Security Agency, "UKUSA Agreement Release, 1940–1956," May 8, 2016.

legal and policy arrangements for intelligence-sharing are already in place, the United States does not always widely release intelligence products to all NATO allies.[7]

To share more information about Russian activities with all NATO allies, the United States would need to dedicate more resources to collecting and analyzing information on Russian activities and rapidly developing and producing NATO-releasable products. In the case of Ukraine, even U.S. national intelligence was reportedly insufficiently informed to provide answers to U.S. policymaker questions, highlighting gaps in regional collection and analysis. Intelligence analysts would need time to write versions of intelligence products that omit the most highly classified information about sources and methods. The United States would also need to increase the authorities of foreign disclosure and release officers who review intelligence products for release to U.S. partners.[8] The dedicated Intelligence Community task force proposed above could take the bureaucratic lead in identifying relevant intelligence relating to Russian activities for release to NATO allies. This task force could assign liaisons to NATO's Assistant Secretary General for Intelligence to speed the flow of information.

Newly shared intelligence on Russian activities could serve as the basis for a broader diplomatic approach to NATO allies. Intelligence-sharing could be particularly critical in the early phases of a conflict with Russia, during which Russian efforts to obscure its activities could influence the perceived threat perception by NATO allies. Expanded intelligence-sharing can significantly contribute to uncovering and countering Russian disinformation campaigns; increased sharing on Russian counterintelligence would also be helpful. Over the longer term, steps to improve intelligence-sharing on Russian motivations and activities could help allies come to a shared view about the nature of the threat posed by Russian aims and activities before any conflict occurred.

Provide More U.S. Personnel to NATO's Intelligence and Analysis Organizations

In light of evolving challenges to NATO's east and south, the alliance has already taken steps to reform its approach to shared intelligence. In addition to creating the JISD, NATO established a special unit within the division to systematically examine unconventional and hybrid threats.[9] The multinational nature of the forum could help legitimize intelligence during an ambiguous scenario. The hybrid unit, in particular, is currently relatively small; additional skilled U.S. personnel could help to expand the scope and depth of the unit's work. As discussed in Chapter Three, the ambiguity about Russian involvement in an unconventional scenario could make it particu-

---

[7] Murray Brewster, "NATO Intelligence Chiefs Admit Sharing Secrets Is a 'Challenge' Within the Alliance," *CBC News*, May 24, 2018.

[8] For a discussion of the role of foreign disclosure and release officers, see Intelligence Community Directive 403, *Foreign Disclosure and Release of Classified National Intelligence*, Washington, D.C.: Office of the Director of National Intelligence, March 13, 2013.

[9] Arndt Freytag von Loringhoven, "Adapting NATO Intelligence in Support of 'One NATO,'" *NATO Review*, September 8, 2017.

larly difficult to agree on the threat that Russia poses. Strengthening the JISD, and its hybrid unit in particular, might be particularly useful for helping allies come to a shared understanding of the facts on the ground.

Relatedly, sending more U.S. personnel to new NATO centers of excellence would provide additional opportunities to develop longer-term ties between analytic communities. For example, the European Centre of Excellence for Countering Hybrid Threats in Helsinki, established in 2017, is meant to serve as a hub of expertise for participating countries. The United States has already joined the organization, but it could send additional personnel to build closer ties with communities working on these issues across the alliance.[10] The United States could also second personnel to the NATO Strategic Communications Centre of Excellence in Riga, which has recently expanded its operations, with eight new sponsoring nations and seconded staff from France and Canada as of May 2018.[11]

Initiate Dedicated Dialogue with Allies on Russian Aims and Motivations

Intelligence-sharing, on its own, might not be sufficient to help allies come to a common view about Russian aims and motivations. Engaging in regular dialogue, including through processes to update allied threat assessments, might also help allies come closer to a shared understanding of the threat Russia poses.

Although NATO members have periodically engaged in such discussions to produce statements at recent summits about cyber, hybrid, and other threats, NATO could benefit from a dedicated discussion on the nature of current security challenges posed by Russia.[12] One initiative that would facilitate a high-level, structured, alliance-wide discussion on Russian aims and motivations would be to update NATO's 2010 Strategic Concept document, which includes a section on the security environment. During the process to create the current document, allies were reportedly divided over how to characterize the emerging challenges posed by Russia.[13] Ultimately, Russia was not explicitly addressed in the Security Environment section: The document makes references to the need for robust defense and deterrence capabilities generally, coupled with a desire to develop a strategic partnership between Russia and NATO.[14] While

---

[10] European Centre of Excellence for Countering Hybrid Threats, homepage, undated.

[11] As of May 2018, Estonia, Germany, Italy, Latvia, Lithuania, Poland, the Netherlands and United Kingdom had signed on as sponsoring nations (NATO Strategic Communications Centre of Excellence, "About Us," webpage, undated).

[12] NATO, 2014a. At the 2016 Warsaw Summit, allies established four multinational battlegroups in Estonia, Latvia, Lithuania, and Poland (NATO, 2016).

[13] "NATO: An Inadequate Strategic Concept?" *Stratfor*, November 22, 2010.

[14] NATO, "Active Engagement, Modern Defence: Strategic Concept for the Defence and Security of the Members of the North Atlantic Treaty Organization Adopted by Heads of State and Government in Lisbon," November 19, 2010.

a renewed discussion on the characterization of the Russian challenge could itself air divisions and vulnerabilities within the alliance, it could also offer important opportunities to engage in a candid discussion with allies about recent and ongoing Russian activities and to seek a consensus position across the alliance.[15]

Supplement Defensive Capabilities of Vulnerable Allies

During an escalating conflict between Russia and NATO, Russia could try to deter allies from making military contributions to collective defense by threatening military strikes against their homelands. It would be technologically infeasible and politically destabilizing to build missile defense systems in Europe that could protect European allies against all Russian missile strikes. As a result, allies will likely have to face the threat of Russian retaliatory missile attacks should they participate in a military response. Still, the United States could encourage allies to take additional steps that reduce their vulnerability and therefore the coercive power of Russia's retaliatory capabilities.

Encourage Allies to Invest in Point and Small-Area Missile Defenses

NATO's missile defense system is not designed to provide territorial defense of Europe or the United States against a Russian ballistic missile threat, nor is it geographically well positioned to do so. Point defense of important assets—including military bases, critical infrastructure, and population centers—against Russian short-range ballistic missiles is theoretically possible, but NATO currently faces capacity and capability shortfalls.[16] Looking ahead, one U.S. capability currently in development—the Indirect Fire Protection Capability Increment 2—could lead to a potentially lower-cost cruise missile defense system for point or small-area defense because of its open (nonproprietary) architecture that accepts different interceptors.[17] The U.S. Army is currently investing in new interceptors for the system; thought should also be given to how indirect fire protection capability could be made available to allies in Europe.

---

[15] For another discussion of the value of a shared threat assessment, see Andrew A. Michta, "A Common Threat Assessment for NATO?" *Judy Dempsey's Strategic Europe*, February 16, 2017.

[16] Allies are currently taking steps to improve their point defenses. Poland signed a letter of offer and acceptance in March 2018 with the U.S. government to buy two Patriot Configuration 3+ batteries, the latest version of the Patriot system; in November 2017, Romania signed a letter of offer and acceptance to buy up to seven of the unit with two fire units per battery (Jen Judson, "Poland Officially Signs Deal to Buy Patriot from U.S.," *DefenseNews*, March 28, 2018b; Jen Judson, "It's Official: Romania Signs Deal to Buy U.S. Missile Defense System," *DefenseNews*, November 29, 2017). Germany has also taken steps to partner with Lockheed, MBDA Deutschland, and Italy's Leonardo to develop the Medium Extended Air Defense System, a move that could potentially offer Europeans additional air and missile defense options in the coming years.

[17] Tamir Eshel, "New Launcher to Deploy C-RAM, C-UAV, and Counter Cruise-Missile Defenses by 2019," *Defense Update*, March 28, 2015; Jen Judson, "U.S. Army Seeks New Missile Counter Drones, Rockets, and More," *DefenseNews*, February 23, 2018a.

Encourage Allies to Conduct Civil Defense Exercises and Preparations

Taking steps to improve civil defense during peacetime could help allies reduce their vulnerability to Russian retaliation in the event of a conflict. Bilaterally or through NATO's Civil Emergency Planning Committee, the United States should work with allies to identify necessary civil preparedness measures in the event of Russian strikes, including cyberattacks. Current NATO civil emergency planning through the Civil Emergency Planning Committee addresses terrorism response, humanitarian and disaster response, and protection of critical infrastructure. Explicitly focusing on Russian contingencies would reflect the evolving security environment, prompt difficult but necessary discussions within national capitals about options in response to a Russian threat, and contribute to a more resilient alliance.

Mitigate Economic Risk to Allies

Allies that do not perceive an existential threat from Russia might be particularly concerned about Russia's ability to retaliate with economic sanctions, especially during an unconventional scenario. Steps to reduce allies' economic dependence on Russia could make them less vulnerable to Russian coercion.

Propose NATO Dialogue with the EU on Diversifying European Energy Supplies

While NATO currently works to protect critical energy infrastructure and consults with member governments to enhance strategic awareness about energy issues, the organization has stated that it will seek to do more in the future.[18] Given European reliance on Russian gas, NATO should prioritize support for frontline countries developing global liquefied gas (LNG) infrastructure (such as Poland and Lithuania), allies developing new interconnectors and energy resources (such as Romania), and pipelines projects that diversify sources of gas, including the Southern Gas Corridor project to bring Azerbaijani gas to Southern Europe. U.S. officials have already publicly addressed the potentially negative ramifications of new Russian gas pipelines into Europe.[19] As an alliance, NATO should enhance its strategic dialogue on energy with the EU to identify ways to mitigate the negative security implications of future Russian pipelines (the Nord Stream 2, which terminates in Germany, and the TurkStream pipeline to Turkey's European territory).

Dependence on Russian energy imports has traditionally made Poland particularly vulnerable to energy pressure.[20] However, Russia's ability to use energy threats as

[18] NATO, "NATO's Role in Energy Security," webpage, September 14, 2018e.

[19] David B. Rivkin, Jr., and Miomir Zuzul, "Trump Is Right on Nord Stream 2," *Wall Street Journal*, July 22, 2018.

[20] International Energy Agency, *Poland*, Paris, undated.

leverage could well decline as Poland and other states in Eastern Europe pursue energy independence through LNG imports and regional connector pipelines.[21]

Consider Joint NATO-EU Strategic Reserves Program

During a period of crisis, allies dependent on Russian gas might be more likely to risk an interruption in their gas supply if strategic reserves are available. NATO and the EU should work together to consider contingency options to provide LNG or other gas through regional interconnectors to an ally under pressure.[22]

Alliance Politics

Consistently Signal U.S. Commitment to NATO

States are likely to consider how decisions to use force in support of an ally would affect their future alliance relationships. If allies believe that failure to support an attacked ally would be catastrophic for the alliance, and particularly for their relationship with the United States, they might be more likely to contribute. In order for this dynamic to be relevant, allies must believe that the United States would otherwise remain committed to the alliance. Thus, strategic messaging from senior U.S. political leaders and diplomats should emphasize the continued priority that the United States places on its transatlantic security relationships; a commitment underscored through senior-level regional visits, active participation in NATO initiatives, and robust U.S. military presence in Europe. The United States should continue to demonstrate, through statements, posture, and activities, its unwavering commitment to NATO's Article 5 security guarantee.

In light of sometimes mixed signals from the U.S. executive branch, bipartisan signaling from the U.S. Congress might be particularly important in communicating U.S. commitment to parliamentary representatives across the alliance. The Helsinki Commission's 2018 resolution "expressing the sense of the Senate regarding the strategic importance of NATO to the collective security of the transatlantic region and urging member states to work together at the upcoming summit to strengthen the alliance" provides one positive example.[23] Clear and consistent U.S. messaging about alliance value would be particularly critical during a crisis, when Russian information campaigns would be likely to be most active.

---

[21] Agnieszka Barteczko, "Poland Buys More LNG, Reduces Reliance on Russian Gas," *Reuters*, August 3, 2018.

[22] Leonardo Maugeri, "Europe Needs a Strategic Gas Reserve," Harvard Kennedy School, Belfer Center for Science and International Affairs, October 2014.

[23] U.S. Senate Committee on Foreign Relations, *A Resolution Expressing the Sense of the Senate Regarding the Strategic Importance of NATO to the Collective Security of the Transatlantic Region and Urging Its Member States to Work Together at the Upcoming Summit to Strengthen the Alliance*, Washington, D.C.: U.S. Government Publishing Office, S. Res. 557, July 17, 2018b.

Communicate Expectations for Participation and Clear Consequences for Failing to Support an Ally Under Attack

In addition to assuring allies of U.S. commitment, U.S. officials should also make clear that all allies would be expected to contribute to collective defense contingencies, and that failure to do so could adversely affect their bilateral relationships with the United States. More controversially, the United States could state explicitly that its security guarantees are conditional upon an ally's willingness to support an Article 5 contingency.

Address Allied Concerns About Entrapment

As discussed in Chapter Two, states are sometimes reluctant to support an ally because of the risk of entrapment (being pulled into conflict by an ally's policy choices). Conversely, states are more likely to be willing to participate in a military response if they believe that other allies are taking steps to avoid conflict and will consult before taking steps that could escalate conflict with Russia. For example, timely use of Article 4 consultations ahead of Article 5 crises can considerably reduce allies' anxieties of being rushed into a conflict that might otherwise be avoided. In this context, allies are most likely to be concerned about provocations by Baltic states or that the United States has more-ambitious aims with respect to Russia.

Encourage Baltic States to Continue to Address Their Russian Populations' Grievances

Encouraging Baltic states to continue outreach to ethnic Russian populations—and to heavily publicize their efforts—might address potential sources of conflict with Russia and reassure NATO allies that Baltic states are fully committed to removing Russian pretexts for intervention.[24]

Relatedly, sustaining support for NATO Counter Hybrid Support Teams could also help the Baltic states feel that they have the tools to manage Russian attempts to influence these populations. These teams are intended to provide robust defensive assistance to states experiencing hybrid aggression, giving them a viable NATO option for support short of conventional escalation.[25]

Encourage Threatened States to Use Article 4 Consultations Early in a Crisis

Article 4 meetings are infrequent, but they can strengthen NATO cohesion in a crisis by offering allies that feel threatened a direct channel to solicit support and by offering allies that fear entrapment an opportunity to identify and review alternative courses

---

[24] The Baltic states have some efforts already underway. For example, Estonia has taken steps to simplify the process for Russian speakers to obtain Estonian citizenship and to improve the lagging economy in communities near the Russian border (Alistair Scrutton and David Mardiste, "Wary of Divided Loyalties, a Baltic State Reaches Out to Its Russians," *Reuters*, February 24, 2017).

[25] NATO, 2018a.

of action. During the initial weeks of the crisis in Ukraine, repeated NATO Article 4 consultations served as a mechanism for allies of Poland to air concerns about Russian regional aggression and for other allies to exchange views on how best to address the crisis.[26] To address concerns about entrapment and reduce the impact of Russian disinformation campaigns blaming an ally for escalation, U.S. leaders should encourage allies to invoke Article 4 and communicate with the NAC immediately during a crisis.

Continue to Promote Dialogue with Russia

The United States, which has at times pursued a more hardline policy toward Russia than other allies, might also need to undertake steps during peacetime to address allied concerns about entrapment. Although NATO suspended some forms of cooperation through the NATO-Russia Council as a result of Russia's incursion into Ukraine in 2014, dialogue on key security concerns has continued. Continued U.S. support for sustaining the NATO-Russia Council as a venue for discussion, even during a crisis, would be one way to show allies that the United States seeks to avoid unnecessary conflict with Russia. New steps to address conflicts of interest with Russia would be an even stronger signal of U.S. commitment to avoid unnecessary conflict.

Provoke Discussion of Political Goals by Using Conventional and Unconventional Russia Scenarios in Crisis Management Exercises

Allies are more likely to contribute to a NATO military operation if they believe that the operation's goals are consistent with their own, and that their voices will be heard during the decisionmaking process. NATO crisis management exercises at the NAC level have recently begun to address Russia scenarios. Future crisis management exercises and other high-level discussions could address a range of plausible Russia scenarios—conventional and unconventional—and focus on the decisions that each ally might need to make about whether and how to contribute. In addition to operational considerations, these fora should be used to address national perspectives on appropriate political and strategic goals during a notional conflict with Russia. Through these exercises, the United States and NATO allies should seek to develop consensus on policy objectives in peacetime in order to facilitate national decisionmaking if a crisis occurs.

Final Thoughts

The framework established in this report provides an analytical tool for planners and policymakers considering the likelihood of a given state's participation in a military response to Russian aggression. We applied this framework to explore how decision-

[26] Justyna Pawlak and Adrian Croft, "Poland Requests More NATO Consultations over Russia," *Reuters*, March 3, 2014.

making might change depending on the nature of a Russian attack, identify ways that Russia could manipulate allies' decisionmaking, and propose policies to reduce allies' vulnerabilities and make military contributions more likely.

The factors likely to drive allies' decisionmaking could vary substantially depending on whether Russia uses conventional or unconventional means. For example, widely discussed proposals to counter Russian influence attempts against the public and policymakers would be particularly important during an unconventional conflict or phase. Because the value that allied political elites place on the perpetuation of NATO is likely to be a determinative factor in a decision to commit forces to a conventional conflict with Russia, investments now to engage emerging political elites on NATO value could have tangible implications for a future conventional fight. Ultimately, U.S. threat assessments should guide U.S. policy priorities for increasing allies' contributions to collective defense.

Our decisionmaking framework has highlighted other policies that have not been as widely considered. Allies' concerns about entrapment could make contributions to collective defense less likely, especially in a conventional scenario in which escalation is more likely. Allies are likely to focus on how U.S. goals align with their own and whether the United States is willing to be restrained within collective decisionmaking structures. In Afghanistan, the United States initially chose to operate outside of restrictive NATO decisionmaking structures. In a conflict with Russia, which is much more militarily capable than recent adversaries, the United States and its NATO allies would rely more heavily on each other than in any conflict since the establishment of the alliance. As a result, the United States could face unprecedented challenges in finding the balance between ensuring its own freedom of action and leading the effort to rally allies to commit their forces. The recommendations in this report aim to help the United States navigate this challenge.

References

Adamkus, Valdas, Martin Butora, Emil Constantinescu, Pavol Demes, Lubos Dubrovsky, Matyas Eorsi, Istvan Gyarmati, Vaclav Havel, Rastislav Kacer, Sandra Kalniete, Karel Schwarzenberg, Michal Kovac, Ivan Krastev, Alexander Kwasniewski, Matt Laar, Kadri Liik, Janos Martonyi, Janusz Onyszkiewicz, Adam Rotfield, Vaira Vike-Freiberga, Alexandr Vondra, and Lech Walesa, "An Open Letter to the Obama Administration from Central and Eastern Europe," *Radio Free Europe/Radio Liberty*, July 16, 2009. As of September 19, 2018:
https://www.rferl.org/a/An_Open_Letter_To_The_Obama_Administration_From_Central_And_Eastern_Europe/1778449.html

AfD—*See* Alternative für Deutschland.

Aldrich, John H., Christopher Gelpi, Peter Feaver, Jason Reifler, and Kristin Thompson Sharp, "Foreign Policy and the Electoral Connection," *Annual Review of Political Science*, Vol. 9, June 2006, pp. 477–502. As of September 17, 2018:
https://www.annualreviews.org/doi/abs/10.1146/annurev.polisci.9.111605.105008

Allen, Michael A., Julie van Dusky-Allen, and Michael E. Flynn, "The Localized and Spatial Effects of U.S. Troop Deployments on Host-State Defense Spending," *Foreign Policy Analysis*, Vol. 12, No. 4, October 2016, pp. 674–694. As of September 19, 2018:
https://academic.oup.com/fpa/article/12/4/674/2469900

Alternative für Deutschland, *Manifesto for Germany: The Political Programme of the Alternative for Germany*, undated. As of October 4, 2018:
https://www.afd.de/wp-content/uploads/sites/111/2017/04/2017-04-12_afd-grundsatzprogramm-englisch_web.pdf

Alvarez, Lizette, and Elaine Sciolino, "Bombings in Madrid: Election Outcome; Spain Grapples with Notion that Terrorism Trumped Democracy," *New York Times*, March 17, 2004. As of September 19, 2018:
https://www.nytimes.com/2004/03/17/world/bombings-madrid-election-outcome-spain-grapples-with-notion-that-terrorism.html

Anischchuk, Alexei, "Putin Admits Russian Forces Were Deployed to Crimea," *Reuters*, April 17, 2014. As of September 20, 2018:
https://www.reuters.com/article/russia-putin-crimea/putin-admits-russian-forces-were-deployed-to-crimea-idUSL6N0N921H20140417

Auerswald, David P., "Inward Bound: Domestic Institutions and Military Conflicts," *International Organization*, Vol. 53, No. 3, Summer, 1999, pp. 469–504. As of September 19, 2018:
http://www.jstor.org/stable/2601287

Aydıntaşbaş, Asli, *With Friends Like These: Turkey, Russia, and the End of an Unlikely Alliance*, London: European Council on Foreign Relations, June 23, 2016. As of September 20, 2018: https://www.ecfr.eu/publications/summary/with_friends_like_these_turkey_russia_and_the_end_of_an_unlikely_7048

Baldwin, David A., "The Sanctions Debate and the Logic of Choice," *International Security*, Vol. 24, No. 3, Winter 1999, pp. 80–107. As of September 19, 2018: http://www.jstor.org/stable/2539306

Barteczko, Agnieszka, "Poland Buys More LNG, Reduces Reliance on Russian Gas," *Reuters*, August 3, 2018. As of September 20, 2018: https://www.reuters.com/article/pgnig-gazprom/poland-buys-more-lng-reduces-reliance-on-russian-gas-idUSL5N1UU618

Batchelor, Tom, "Marine Le Pen Insists Russian Annexation of Crimea Is Totally Legitimate," *The Independent*, January 3, 2017. As of September 19, 2018: https://www.independent.co.uk/news/world/europe/marine-le-pen-crimea-russia-putin-ukraine-illegal-annexation-france-front-national-fn-a7507361.html

Baum, Matthew A., and Philip B. K. Potter, "The Relationships Between Mass Media, Public Opinion, and Foreign Policy: Toward a Theoretical Synthesis," *Annual Review of Political Science*, Vol. 11, June 2008, pp. 39–65. As of September 17, 2018: https://www.annualreviews.org/doi/full/10.1146/annurev.polisci.11.060406.214132

Beckley, Michael, "The Myth of Entangling Alliances: Reassessing the Security Risks of U.S. Defense Pacts," *International Security*, Vol. 39, No. 4, Spring, 2015, pp. 7–48. As of September 19, 2018: https://www.mitpressjournals.org/doi/pdf/10.1162/ISEC_a_00197

Bennett, Andrew, Joseph Lepgold, and Danny Unger, "Burden-Sharing in the Persian Gulf War," *International Organization*, Vol. 48, No. 1, Winter 1994, pp. 39–75. As of September 17, 2018: http://www.jstor.org/stable/2706914

Berinsky, Adam J., "Assuming the Costs of War: Events, Elites, and American Public Support for Military Conflict," *Journal of Politics*, Vol. 69, No. 4, November, 2007, pp. 975–997. As of September 19, 2018: http://www.jstor.org/stable/10.1111/j.1468-2508.2007.00602.x

Blank, Stephen, "Putin's Next Potential Target: The Baltic States," *RealClear Defense*, January 5, 2016. As of September 19, 2018: https://www.realcleardefense.com/articles/2016/01/06/putins_next_potential_target_the_baltic_states_108864.html

Bohush, Denis, and Oleksandra Baglai, *Russian Propaganda Concerning Ukraine During the Syrian Campaign: An Innovative Approach to Assess Information Activities*, Riga, Latvia: NATO Strategic Communications Centre of Excellence, November 2016. As of September 20, 2018: https://www.stratcomcoe.org/russian-propaganda-concerning-ukraine-during-syrian-campaign-innovative-approach-assess-information

Bozo, Frédéric, "Explaining France's NATO 'Normalisation' Under Nicolas Sarkozy (2007–2012)," *Journal of Transatlantic Studies*, Vol. 12, No. 4, December, 2014, pp. 379–391. As of September 19, 2018: https://www.tandfonline.com/doi/pdf/10.1080/14794012.2014.962737?needAccess=true

"Breedlove's Bellicosity: Berlin Alarmed by Aggressive NATO Stance on Ukraine," *Spiegel Online*, March 6, 2015. As of September 20, 2018: http://www.spiegel.de/international/world/germany-concerned-about-aggressive-nato-stance-on-ukraine-a-1022193.html

Bremer, Stuart A., "Dangerous Dyads: Conditions Affecting the Likelihood of Interstate War, 1816–1965," *Journal of Conflict Resolution*, Vol. 36, No. 2, June, 1992, pp. 309–341. As of February 18, 2019:
http://www.jstor.org/stable/174478

Brewster, Murray, "NATO Intelligence Chiefs Admit Sharing Secrets Is a 'Challenge' Within the Alliance," *CBC News*, May 24, 2018. As of September 20, 2018:
https://www.cbc.ca/news/politics/nato-intelligence-chiefs-meeting-1.4677045

Brooke-Holland, Louisa, *NATO's Military Response to Russia: November 2016 Update*, London: House of Commons Library, Briefing Paper No. 07276, November 3, 2016. As of September 17, 2018:
https://researchbriefings.parliament.uk/ResearchBriefing/Summary/CBP-7276#fullreport

"Bulgaria Says Will Not Join Any NATO Black Sea Fleet After Russian Warning," *Reuters*, June 16, 2016. As of September 19, 2018:
https://www.reuters.com/article/nato-bulgaria-blacksea-idUSL8N19835X

Buras, Piotr, and Adam Balcer, "An Unpredictable Russia: The Impact on Poland," European Council on Foreign Relations, July 15, 2016. As of September 19, 2018:
https://www.ecfr.eu/article/commentary_an_unpredictable_russia_the_impact_on_poland

Cameron, David, *National Security Strategy and Strategic Defence and Security Review 2015: A Secure and Prosperous United Kingdom*, London, U.K.: Her Majesty's Stationery Office, CM9161, November 2015. As of September 19, 2018:
https://assets.publishing.service.gov.uk/government/uploads/system/uploads/attachment_data/file/555607/2015_Strategic_Defence_and_Security_Review.pdf

Chappell, Bill, and Mark Memmott, "Putin Says Those Aren't Russian Forces in Crimea," *NPR*, March 4, 2014. As of September 19, 2018:
https://www.npr.org/sections/thetwo-way/2014/03/04/285653335/putin-says-those-arent-russian-forces-in-crimea

"Chapter Four: Europe," *Military Balance*, Vol. 117, No. 1, 2017, pp. 63–182. As of September 19, 2018:
https://www.tandfonline.com/doi/abs/10.1080/04597222.2017.1271210

Chee, Foo Yun, "NATO Says It Sees Sharp Rise in Russian Disinformation Since Crimea Seizure," *Reuters*, February 11, 2017. As of September 20, 2018:
https://www.reuters.com/article/us-ukraine-crisis-russia-media/nato-says-it-sees-sharp-rise-in-russian-disinformation-since-crimea-seizure-idUSKBN15Q0MG

Christensen, Thomas J., *Worse Than a Monolith: Alliance Politics and Problems of Coercive Diplomacy in Asia*, Princeton, N.J.: Princeton University Press, 2011.

Chuck, Elizabeth, "White House to Lay out U.S. Intelligence on MH17," *NBC News*, July 22, 2014. As of September 20, 2018:
https://www.nbcnews.com/storyline/ukraine-plane-crash/white-house-lay-out-u-s-intelligence-mh17-n162236

Clark, Wesley, Jüri Luik, Egon Ramms, and Richard Shirreff, *Closing NATO's Baltic Gap*, Tallinn, Estonia: International Centre for Defence and Security, May 2016.

Coalson, Robert, "What Are NATO's Articles 4 and 5?" *Radio Free Europe/Radio Liberty*, June 26, 2012. As of September 17, 2018:
https://www.rferl.org/a/explainer-nato-articles-4-and-5/24626653.html

Cohen, Ariel, and Robert E. Hamilton, *The Russian Military and the Georgia War: Lessons and Implications*, Carlisle Barracks, Pa.: Strategic Studies Institute, U.S. Army War College, June 2011. As of September 20, 2018:
http://ssi.armywarcollege.edu/pdffiles/pub1069.pdf

Daley, Suzanne, "After the Attacks: The Alliance; For First Time, NATO Invokes Joint Defense Pact with U.S.," *New York Times*, September 13, 2001. As of January 8, 2019:
https://www.nytimes.com/2001/09/13/us/after-attacks-alliance-for-first-time-nato-invokes-joint-defense-pact-with-us.html

Davidson, Jason W., *America's Allies and War: Kosovo, Afghanistan, and Iraq*, New York: Palgrave Macmillan, 2011.

Defense Intelligence Agency, *Russia Military Power: Building a Military to Support Great Power Aspirations*, Washington, D.C., DIA-11-1704-161, 2017. As of September 19, 2018:
http://www.dia.mil/Portals/27/Documents/News/Military%20Power%20Publications/Russia%20Military%20Power%20Report%202017.pdf

de Graaf, Beatrice, George Dimitriu, and Jens Ringsmose, "Conclusion—How to Operate Strategic Narratives: Interweaving War, Politics, and the Public," in Beatrice de Graaf, George Dimitriu, and Jens Ringsmose, eds., *Strategic Narratives, Public Opinion, and War: Winning Domestic Support for the Afghan War*, New York: Routledge, 2015.

Demirjian, Karoun, and Michael Birnbaum, "Russia Escalates Tensions with Aid Convoy, Reported Firing of Artillery Inside Ukraine," *Washington Post*, July 22, 2017. As of September 20, 2018:
https://www.washingtonpost.com/world/russian-humanitarian-convoy-enters-ukraine-despite-warnings/2014/08/22/7b14fa8e-29e1-11e4-8593-da634b334390_story.html?utm_term=.41b71fa7fdda

Dempsey, Judy, "NATO's European Allies Won't Fight for Article 5," *Judy Dempsey's Strategic Europe*, June 15, 2015. As of September 17, 2018:
http://carnegieeurope.eu/strategiceurope/?fa=60389

Dougherty, Jill, and Riina Kaljurand, *Estonia's "Virtual Russian World": The Influence of Russian Media on Estonia's Russian Speakers*, Tallinn, Estonia: International Centre for Defence and Security, October 2015.

Drent, Margriet, Peter van Ham, and Kees Homan, *Article 5 Revisited—Is NATO Up to It?* The Hague, Netherlands: Clingendael, August 2014. As of September 17, 2018:
https://www.clingendael.org/sites/default/files/pdfs/Article-5-revisited-is-NATO-up-to-it.pdf

Dyomkin, Denis, "Putin Says Romania, Poland May Now Be in Russia's Cross-Hairs," *Reuters*, May 27, 2016. As of September 19, 2018:
https://www.reuters.com/article/us-russia-europe-shield-idUSKCN0YI2ER

Eichenberg, Richard C., "Victory Has Many Friends: U.S. Public Opinion and the Use of Military Force, 1981–2005," *International Security*, Vol. 30, No. 1, Summer 2005, pp. 140–177. As of September 19, 2018:
http://www.jstor.org/stable/4137461

Eshel, Tamir, "New Launcher to Deploy C-RAM, C-UAV, and Counter Cruise-Missile Defenses by 2019," *Defense Update*, March 28, 2015. As of September 20, 2018:
https://defense-update.com/20150328_mml.html

European Centre of Excellence for Countering Hybrid Threats, homepage, undated. As of September 20, 2018:
https://www.hybridcoe.fi

Fang, Songying, Jesse C. Johnson, and Brett Ashley Leeds, "To Concede or to Resist? The Restraining Effect of Military Alliances," *International Organization*, Vol. 68, No. 4, Fall 2014, pp. 775–809. As of September 19, 2018:
https://www.cambridge.org/core/services/aop-cambridge-core/content/view/5AE6E043122ECF3754F42E106362DECC/S0020818314000137a.pdf/to_concede_or_to_resist_the_restraining_effect_of_military_alliances.pdf

Fearon, James D., "Domestic Political Audiences and the Escalation of International Disputes," *American Political Science Review*, Vol. 88, No. 3, September 1994, pp. 577–592. As of September 19, 2018:
https://www.jstor.org/stable/2944796?seq=1#metadata_info_tab_contents

Federal Ministry of Defence, *White Paper 2016: On German Security Policy and the Future of the Bundeswehr*, Berlin: Federal Ministry of Defence, Federal Government of Germany, July 13, 2016.

Fletcher School, Tufts University, "SIMULEX 2016: Crisis in the Baltic Region and the Middle East," crisis management exercise, October 21–22, 2016.

Foley, James B., "Don't Let Putin Destroy NATO," *Time*, March 31, 2016. As of September 17, 2018:
http://time.com/4276525/vladimir-putin-nato

Freedman, Lawrence, "The Possibilities and Limits of Strategic Narratives," in Beatrice de Graaf, George Dimitriu, and Jens Ringsmose, eds., *Strategic Narratives, Public Opinion, and War: Winning Domestic Support for the Afghan War*, New York: Routledge, 2015.

Garčević, Vesko, "Russian Interference in European Elections," hearing before the U.S. Senate Select Committee on Intelligence, Washington, D.C., June 28, 2017. As of April 19, 2019:
https://www.intelligence.senate.gov/sites/default/files/documents/sfr-vgarcevic-062817b.pdf

Gelpi, Christopher, and Joseph M. Grieco, "Competency Costs in Foreign Affairs: Presidential Performance in International Conflicts and Domestic Legislative Success, 1953–2001," *American Journal of Political Science*, Vol. 59, No. 2, April 2015, pp. 440–456. As of September 19, 2018:
http://dx.doi.org/10.1111/ajps.12169

"Germany's Libya Contribution: Merkel Approves AWACS for Afghanistan," *Spiegel Online*, March 23, 2011. As of September 19, 2018:
http://www.spiegel.de/international/world/germany-s-libya-contribution-merkel-cabinet-approves-awacs-for-afghanistan-a-752709.html

Gerzhoy, Gene, "Alliance Coercion and Nuclear Restraint: How the United States Thwarted West Germany's Nuclear Ambitions," *International Security*, Vol. 39, No. 4, Spring 2015, pp. 91–129. As of September 19, 2018:
http://dx.doi.org/10.1162/ISEC_a_00198

Giles, Keir, *Russia's "New" Tools for Confronting the West: Continuity and Innovation in Moscow's Exercise of Power*, London: Chatham House, Royal Institute of International Affairs, March 2016. As of September 21, 2018:
https://www.chathamhouse.org/sites/default/files/publications/2016-03-russia-new-tools-giles.pdf

Goble, Paul A., "Russian 'Passportization,'" *New York Times*, September 9, 2008. As of September 9, 2018:
https://topics.blogs.nytimes.com/2008/09/09/russian-passportization/?_r=2&mtrref=undefined&gwh=0AAAD60787B2316B5F4494AE0A5BDD8&gwt=pay

Gordon, Michael R., and Niraj Chokshi, "Trump Criticizes NATO and Hopes for 'Good Deals' With Russia," *New York Times*, January 15, 2017. As of September 17, 2018:
https://www.nytimes.com/2017/01/15/world/europe/donald-trump-nato.html

Gordon, Philip H., and Jeremy Shapiro, *Allies at War: America, Europe, and the Crisis over Iraq*, Washington, D.C.: Brookings Institution Press, 2004.

Helmus, Todd C., Elizabeth Bodine-Baron, Andrew Radin, Madeline Magnuson, Joshua Mendelsohn, William Marcellino, Andriy Bega, and Zev Winkelman, *Russian Social Media Influence: Understanding Russian Propaganda in Eastern Europe*, Santa Monica, Calif.: RAND Corporation, RR-2237-OSD, 2018. As of September 19, 2018:
https://www.rand.org/pubs/research_reports/RR2237.html

Herszenhorn, David M., "Satellites Show Russian Forces Poised Near Ukraine," *New York Times*, April 10, 2014. As of October 4, 2018:
https://www.nytimes.com/2014/04/11/world/europe/satellites-show-russia-mobilizing-near-ukraine-nato-says.html

———, "Putin Extends Counter-Sanctions Against EU," *Politico*, June 30, 2017. As of September 20, 2018:
https://www.politico.eu/article/putin-extends-counter-sanctions-against-eu

Herszenhorn, David M., and Lili Bayer, "Trump's Whiplash NATO Summit," *Politico*, July 12, 2018. As of January 8, 2019:
https://www.politico.eu/article/trump-threatens-to-pull-out-of-nato

Hille, Kathrin, "Russia Attacks NATO's Decision to Strengthen Defences in Poland," *Financial Times*, July 10, 2016. As of September 19, 2018:
https://www.ft.com/content/8b6092a0-46a8-11e6-8d68-72e9211e86ab

Hoffmann, Christiane, and René Pfister, "Part of the West? German Leftists Have Still Not Understood Putin—Interview with Historian Henrich Winkler About Germany and the West," *Spiegel Online*, June 27, 2014. As of September 17, 2018:
http://www.spiegel.de/international/germany/interview-with-historian-heinrich-winkler-about-germany-and-the-west-a-977649.html

Hooper, John, "German Leader Says No to Iraq War," *The Guardian*, August 5, 2002. As of September 19, 2018:
https://www.theguardian.com/world/2002/aug/06/iraq.johnhooper

Howard, Michael, *The Causes of War and Other Essays*, Cambridge, Mass.: Harvard University Press, 1983.

Hufbauer, Gary Clyde, Jeffrey J. Schott, and Kimberly Ann Elliott, *Economic Sanctions Reconsidered: History and Current Policy*, Washington, D.C.: Institute for International Economics, 1990.

Hunter, Robert E., "NATO's Article 5: The Conditions for a Military and Political Coalition," *European Affairs*, Vol. 2, No. 4, Fall 2001.

Ikenberry, G. John, *After Victory: Institutions, Strategic Restraint, and the Rebuilding of Order After Major Wars*, Princeton, N.J.: Princeton University Press, 2001.

Intelligence Community Directive 403, *Foreign Disclosure and Release of Classified National Intelligence*, Washington, D.C.: Office of the Director of National Intelligence, March 13, 2013. As of September 20, 2018:
https://www.dni.gov/files/documents/ICD/ICD403.pdf

International Energy Agency, *Poland*, Paris, undated. As of March 18, 2019:
https://cnpp.iaea.org/countryprofiles/Poland/Poland.htm

"Invocation of Article 5 Confirmed," *NATO Update*, October 3, 2001. As of September 17, 2018:
https://www.nato.int/docu/update/2001/1001/e1002a.htm

Jackson, Mike, "Gen Sir Mike Jackson: My Clash with NATO Chief," *The Telegraph*, September 4, 2007. As of September 19, 2018:
https://www.telegraph.co.uk/news/worldnews/1562161/
Gen-Sir-Mike-Jackson-My-clash-with-Nato-chief.html

Janda, Jakub, "How Czech President Miloš Zeman Became Putin's Man," *The Observer*, January 26, 2018. As of September 19, 2018:
https://observer.com/2018/01/how-czech-president-milos-zeman-became-vladimir-putins-man

Janda, Jakub, and Veronika Víchová, "The Kremlin's Hostile Influence in the Czech Republic: The State of Play," *Warsaw Institute Review*, August 10, 2017. As of September 19, 2018:
https://warsawinstitute.org/kremlins-hostile-influence-czech-republic-state-play

Jentleson, Bruce W., "The Pretty Prudent Public: Post Post-Vietnam American Opinion on the Use of Military Force," *International Studies Quarterly*, Vol. 36, No. 1, March 1992, pp. 49–73. As of September 19, 2018:
http://www.jstor.org/stable/2600916

Jervis, Robert, *Perception and Misperception in International Politics*, Princeton, N.J.: Princeton University Press, 1976.

Jones, Sam, "Estonia Ready to Deal with Russia's 'Little Green Men,'" *Financial Times*, May 13, 2015. As of September 19, 2018:
https://www.ft.com/content/03c5ebde-f95a-11e4-ae65-00144feab7de

Jordans, Frank, "European Spy Chiefs Warn of Hybrid Threats from Russia, ISIS," *Fifth Domain*, May 14, 2018. As of February 18, 2019:
https://www.fifthdomain.com/international/2018/05/14/
european-spy-chiefs-warn-of-hybrid-threats-from-russia-isis/

Jozwiak, Rikaard, "European Commission to Call out Russia for 'Information Warfare,'" *Radio Free Europe/Radio Liberty*, April 25, 2018. As of September 19, 2018:
https://www.rferl.org/a/european-commission-to-call-out-russia-for-information-warfare-/
29192052.html

Judson, Jen, "It's Official: Romania Signs Deal to Buy U.S. Missile Defense System," *DefenseNews*, November 29, 2017. As of September 20, 2018:
https://www.defensenews.com/land/2017/11/30/
its-official-romania-signs-deal-to-buy-us-missile-defense-system

———, "U.S. Army Seeks New Missile Counter Drones, Rockets, and More," *DefenseNews*, February 23, 2018a. As of September 20, 2018:
https://www.defensenews.com/land/2018/02/23/
army-wants-brand-new-missile-to-counter-wide-variety-of-threats

———, "Poland Officially Signs Deal to Buy Patriot from U.S.," *DefenseNews*, March 28, 2018b. As of September 20, 2018:
https://www.defensenews.com/land/2018/03/28/poland-officially-signs-deal-to-buy-patriot-from-us

Karnitschnig, Matthew, "NATO's Germany Problem," *Politico*, July 8, 2016. As of September 19, 2018:
https://www.politico.eu/article/
natos-germany-russia-problem-german-foreign-minister-frank-walter-steinmeier

Keohane, Robert O., *After Hegemony: Cooperation and Discord in the World Political Economy*, Princeton, N.J.: Princeton University Press, 1984.

Kepe, Marta, "NATO: Prepared for Countering Disinformation Operations in the Baltic States?" *RAND Blog*, June 7, 2017. As of September 19, 2018:
https://www.rand.org/blog/2017/06/nato-prepared-for-countering-disinformation-operations.html

Kim, Tongfi, "Why Alliances Entangle but Seldom Entrap States," *Security Studies*, Vol. 20, No. 3, August 2011, pp. 350–377. As of September 19, 2018:
http://dx.doi.org/10.1080/09636412.2011.599201

Kington, Tom, "Future of F-35 in Italy Remains a Mystery Under New Government," *DefenseNews*, May 25, 2018. As of October 4, 2018:
https://www.defensenews.com/global/europe/2018/05/25/
future-of-f-35-in-italy-remains-a-mystery-under-new-government

Kofman, Michael, Katya Migacheva, Brian Nichiporuk, Andrew Radin, Olesya Tkacheva, and Jenny Oberholtzer, *Lessons from Russia's Operations in Crimea and Eastern Ukraine*, Santa Monica, Calif.: RAND Corporation, RR-1498-A, 2017. As of September 21, 2018:
https://www.rand.org/pubs/research_reports/RR1498.html

Kremlin Watch Program, *Kremlin Influence in Visegrad Countries and Romania: Overview of the Threat, Existing Countermeasures, and Recommended Next Steps*, Prague, Czech Republic: Wilfried Martens Centre for European Studies and European Values Think-Tank, 2017. As of October 4, 2018:
https://www.europeanvalues.net/wp-content/uploads/2017/12/
Kremlin-Influence-in-Visegrad-Countries-and-Romania.pdf

Kreps, Sarah, "When Does the Mission Determine the Coalition? The Logic of Multilateral Intervention and the Case of Afghanistan," *Security Studies*, Vol. 17, No. 3, September 2008, pp. 531–567. As of September 19, 2018:
https://doi.org/10.1080/09636410802319610

———, "Elite Consensus as a Determinant of Alliance Cohesion: Why Public Opinion Hardly Matters for NATO-Led Operations in Afghanistan," *Foreign Policy Analysis*, Vol. 6, No. 3, July 2010, pp. 191–215. As of September 19, 2018:
http://dx.doi.org/10.1111/j.1743-8594.2010.00108.x

Labs, Eric J., "Beyond Victory: Offensive Realism and the Expansion of War Aims," *Security Studies*, Vol. 6, No. 4, Summer 1997, pp. 1–49. As of September 19, 2018:
http://dx.doi.org/10.1080/09636419708429321

Lake, David A., *Entangling Relations: American Foreign Policy in Its Century*, Princeton, N.J.: Princeton University Press, 1999.

Larson, Eric V., *Casualties and Consensus: The Historical Role of Casualties in Domestic Support for U.S. Military Operations*, Santa Monica, Calif.: RAND Corporation, MR-726-RC, 1996. As of September 19, 2018:
https://www.rand.org/pubs/monograph_reports/MR726.html

"Lavrov Blames NATO for Russian Aggression in Ukraine," *UNIAN*, February 19, 2018. As of September 20, 2018:
https://www.unian.info/politics/10012331-lavrov-blames-nato-for-russian-aggression-in-ukraine.html

Leviev-Sawyer, Clive, "Bulgarian National Security Report Naming Russia as Threat Causes Storm in Parliament," *Sofia Globe*, September 13, 2017. As of October 4, 2018:
https://sofiaglobe.com/2017/09/13/
bulgarian-national-security-report-naming-russia-as-threat-causes-storm-in-parliament/

———, "Change of Stance as Cabinet Adopts Report That Does Not See Russia as Threat to EU, Bulgaria," *Independent Balkan News Agency*, July 23, 2018. As of October 4, 2018:
https://www.balkaneu.com/
change-of-stance-as-cabinet-adopts-report-that-does-not-see-russia-as-threat-to-eu-bulgaria/

Limnéll, Jarno, and Charly Salonius-Pasternak, *Challenge for NATO—Cyber Article 5*, Stockholm, Sweden: Center for Asymmetric Threat Studies, Swedish Defense University, briefing paper, June 2016. As of September 19, 2018:
http://www.diva-portal.org/smash/get/diva2:1119569/FULLTEXT01.pdf

"Lithuanian and Polish Presidents Call for NATO Treaty Article 4 Consultations," *EuroDialogueXXI*, March 3, 2014. As of September 17, 2018:
http://eurodialogue.eu/index.php/
lithuanian-and-polish-presidents-call-nato-treaty-article-4-consultations

"Lithuanian Diplomats Tell Poland: Vilnius Ready to Defend Warsaw," *LETA*, September 27, 2016. As of September 19, 2018:
http://www.leta.lv/eng/defence_matters_eng/defence_matters_eng/news/
CCADC839-E91A-49B9-83E9-8791E2780200/?print

Lostumbo, Michael J., Michael J. McNerney, Eric Peltz, Derek Eaton, David R. Frelinger, Victoria A. Greenfield, John Halliday, Patrick Mills, Bruce R. Nardulli, Stacie L. Pettyjohn, Jerry M. Sollinger, and Stephen M. Worman, *Overseas Basing of U.S. Military Forces: An Assessment of Relative Costs and Strategic Benefits*, Santa Monica, Calif.: RAND Corporation, RR-201-OSD, 2013. As of September 19, 2018:
https://www.rand.org/pubs/research_reports/RR201.html

Lucas, Edward, *The Coming Storm: Baltic Sea Security Report*, Washington, D.C.: Center for European Policy Analysis, June 2015. As of September 19, 2018:
https://www.cepa.org/the-coming-storm

MacFarquhar, Neil, "A Powerful Russian Weapon: The Spread of False Stories," *New York Times*, August 28, 2016. As of September 20, 2018:
https://www.nytimes.com/2016/08/29/world/europe/russia-sweden-disinformation.html

Maness, Ryan C., and Brandon Valeriano, *Russia's Coercive Diplomacy: Energy, Cyber, and Maritime Policy as New Sources of Power*, Basingstoke, U.K.: Palgrave Macmillan UK, 2015.

March, James G., and Johan P. Olsen, "The Institutional Dynamics of International Political Orders," *International Organization*, Vol. 52, No. 4, Autumn 1998, pp. 943–969. As of September 19, 2018:
http://www.jstor.org/stable/2601363

Marrone, Alessandro, and Michele Nones, *Italy and Security in the Mediterranean*, Rome: Instituto Affari Internazionali, 2016. As of September 19, 2018:
http://www.iai.it/sites/default/files/iairp_24.pdf

Martin, Lisa L., "Credibility, Costs, and Institutions: Cooperation on Economic Sanctions," *World Politics*, Vol. 45, No. 3, April, 1993, pp. 406–432. As of April 30, 2019:
http://www.jstor.org/stable/2950724

Matláry, Janne Haaland, and Magnus Petersson, "Introduction: Will Europe Lead in NATO?" in Janne Haaland Matláry and Magnus Petersson, eds., *NATO's European Allies: Military Capability and Political Will*, New York: Palgrave Macmillan, 2013.

Mattis, Jim, *Summary of the 2018 National Defense Strategy of the United States of America: Sharpening the American Military's Competitive Edge*, Washington D.C.: U.S. Department of Defense, 2018. As of August 16, 2018:
https://www.defense.gov/Portals/1/Documents/pubs/2018-National-Defense-Strategy-Summary.pdf

Mattox, Gale A., "Germany: The Legacy of the War in Afghanistan," in Gale A. Mattox and Stephen M. Grenier, eds., *Coalition Challenges in Afghanistan: The Politics of Alliance*, Stanford, Calif.: Stanford University Press, 2015.

Maugeri, Leonardo, "Europe Needs a Strategic Gas Reserve," Harvard Kennedy School, Belfer Center for Science and International Affairs, October 2014. As of September 20, 2018:
https://www.belfercenter.org/publication/europe-needs-strategic-gas-reserve

Mazarr, Michael J., Miranda Priebe, Andrew Radin, and Astrid Stuth Cevallos, *Understanding the Current International Order*, Santa Monica, Calif.: RAND Corporation, RR-1598-OSD, 2016. As of September 19, 2018:
https://www.rand.org/pubs/research_reports/RR1598.html

McBride, James, "What to Know About Italy's 2018 Elections," Council on Foreign Relations, February 14, 2018. As of October 4, 2018:
https://www.cfr.org/backgrounder/what-know-about-italys-2018-elections

McLean, Elena V., and Taehee Whang, "Friends or Foes? Major Trading Partners and the Success of Economic Sanctions," *International Studies Quarterly*, Vol. 54, No. 2, June, 2010, pp. 427–447. As of September 19, 2018:
http://dx.doi.org/10.1111/j.1468-2478.2010.00594.x

McNerney, Michael J., Ben Connable, S. Rebecca Zimmerman, Natasha Lander, Marek N. Posard, Jason J. Castillo, Dan Madden, Ilana Blum, Aaron Frank, Benjamin J. Fernandes, In Hyo Seol, Christopher Paul, and Andrew Parasiliti, *National Will to Fight: Why Some States Keep Fighting and Others Don't*, Santa Monica, Calif.: RAND Corporation, RR-2477-A, 2018. As of February 18, 2019:
https://www.rand.org/pubs/research_reports/RR2477.html

Mearsheimer, John J., "Why the Ukraine Crisis Is the West's Fault: The Liberal Delusions That Provoked Putin," *Foreign Affairs*, Vol. 93, No. 5, September–October 2014, pp. 77–89. As of September 19, 2018:
https://www.foreignaffairs.com/articles/russia-fsu/2014-08-18/why-ukraine-crisis-west-s-fault

Michta, Andrew A., "A Common Threat Assessment for NATO?" *Judy Dempsey's Strategic Europe*, February 16, 2017. As of September 20, 2018:
http://carnegieeurope.eu/strategiceurope/68017

Miller, Nicholas L., "The Secret Success of Nonproliferation Sanctions," *International Organization*, Vol. 68, No. 4, Fall 2014, pp. 913–944. As of September 19, 2018:
https://www.cambridge.org/core/services/aop-cambridge-core/content/view/D0090E1163F6962CAD93BFF45A0C7C62/S0020818314000216a.pdf/secret_success_of_nonproliferation_sanctions.pdf

Miller, Paul D., "How World War III Could Begin in Latvia," *Foreign Policy*, November 16, 2016. As of September 19, 2018:
http://foreignpolicy.com/2016/11/16/how-world-war-iii-could-begin-in-latvia

Moran, Michael, "Turkey's Article 5 Argument Finds No Takers," *Carnegie Corporation of New York*, February 24, 2016. As of September 17, 2018:
https://www.carnegie.org/news/articles/turkeys-article-5-argument-finds-no-takers

Morrow, James D., "Alliances and Asymmetry: An Alternative to the Capability Aggregation Model of Alliances," *American Journal of Political Science*, Vol. 35, No. 4, November 1991, pp. 904–933. As of September 19, 2018:
http://www.jstor.org/stable/2111499

———, "Alliances: Why Write Them Down?" *Annual Review of Political Science*, Vol. 3, June, 2000, pp. 63–83. As of September 19, 2018:
http://www.annualreviews.org/doi/abs/10.1146/annurev.polisci.3.1.63

National Security Agency, "UKUSA Agreement Release, 1940–1956," May 8, 2016. As of September 20, 2018:
https://www.nsa.gov/news-features/declassified-documents/ukusa

NATO—*See* North Atlantic Treaty Organization.

"NATO: An Inadequate Strategic Concept?" *Stratfor*, November 22, 2010. As of September 20, 2018:
https://worldview.stratfor.com/article/nato-inadequate-strategic-concept

"NATO Eastern Flank Members Pledge Closer Ties, Citing Russia," *Associated Press*, June 8, 2018. As of September 19, 2018:
https://apnews.com/dc32cbeb67d1412e838ceb8304b69a96

North Atlantic Treaty Organization, North Atlantic Treaty, Washington, D.C.: North Atlantic Treaty Organization, April 4, 1949. As of September 17, 2018:
http://www.nato.int/cps/en/natohq/official_texts_17120.htm

———, "Active Engagement, Modern Defence: Strategic Concept for the Defence and Security of the Members of the North Atlantic Treaty Organization Adopted by Heads of State and Government in Lisbon," November 19, 2010. As of September 20, 2018:
https://www.nato.int/cps/en/natohq/official_texts_68580.htm

———, *NATO Wales Summit Guide*, September 3, 2014a. As of September 17, 2018:
http://www.nato.int/nato_static_fl2014/assets/pdf/pdf_publications/20141008_140108-SummitGuideWales2014-eng.pdf

———, "Wales Summit Declaration," transcript, September 5, 2014b. As of September 19, 2018:
https://www.nato.int/cps/en/natohq/official_texts_112964.htm

———, "Warsaw Summit Communique," transcript, July 9, 2016. As of September 17, 2018:
http://www.nato.int/cps/en/natohq/official_texts_133169.htm

———, "NATO's Enhanced Forward Presence," fact sheet, February 2018a. As of September 17, 2018:
https://www.nato.int/nato_static_fl2014/assets/pdf/pdf_2018_02/20180213_1802-factsheet-efp.pdf

———, "Collective Defence—Article 5," webpage, June 12, 2018b. As of September 17, 2018:
https://www.nato.int/cps/en/natohq/topics_110496.htm

———, "Brussels Summit Declaration," transcript, July 11, 2018c. As of September 20, 2018:
https://www.nato.int/cps/en/natohq/official_texts_156624.htm

———, "NATO-Russia Relations: The Facts," webpage, September 7, 2018d. As of September 20, 2018:
https://www.nato.int/cps/en/natohq/topics_111767.htm

———, "NATO's Role in Energy Security," webpage, September 14, 2018e. As of September 20, 2018:
https://www.nato.int/cps/ic/natohq/topics_49208.htm

North Atlantic Treaty Organization Strategic Communications Centre of Excellence, "About Us," webpage, undated. As of September 20, 2018:
https://www.stratcomcoe.org/about-us

Office of the Director of National Intelligence, *Assessing Russian Activities and Intentions in Recent U.S. Elections*, Washington, D.C., January 6, 2017. As of April 22, 2019:
https://www.dni.gov/files/documents/ICA_2017_01.pdf.

Oliker, Olga, *Russia's Nuclear Doctrine: What We Know, What We Don't, and What That Means*, Washington, D.C.: Center for Strategic and International Studies, May 2016. As of October 4, 2018:
https://csis-prod.s3.amazonaws.com/s3fs-public/publication/160504_Oliker_RussiasNuclearDoctrine_Web.pdf

Olson, Mancur, Jr., and Richard Zeckhauser, "An Economic Theory of Alliances," *Review of Economics and Statistics*, Vol. 48, No. 3, August, 1966, pp. 266–279. As of September 19, 2018:
https://www.jstor.org/stable/1927082?seq=1#metadata_info_tab_contents

Opperman, Kai, and Henrike Viehrig, "The Public Salience of Foreign and Security Policy in Britain, Germany and France," *West European Politics*, Vol. 32, No. 5, September 2009, pp. 925–942. As of September 19, 2018:
https://www.tandfonline.com/doi/abs/10.1080/01402380903064804

Paterson, Tony, Peter Foster, and Bruno Waterfield, "Angela Merkel: Russia 'Will Not Get Away' with Annexation of Crimea," *The Telegraph*, March 12, 2014. As of October 4, 2018:
https://www.telegraph.co.uk/news/worldnews/europe/ukraine/10693400/Angela-Merkel-Russia-will-not-get-away-with-annexation-of-Crimea.html

Pawlak, Justyna, and Adrian Croft, "Poland Requests More NATO Consultations over Russia," *Reuters*, March 3, 2014. As of October 4, 2018:
https://www.reuters.com/article/us-ukraine-crisis-nato-meeting/poland-requests-more-nato-consultations-over-russia-idUSBREA221VS20140303

Pezard, Stephanie, "The Front National and the Future of French Foreign Policy," *War on the Rocks*, March 25, 2015. As of October 4, 2018:
https://warontherocks.com/2015/03/the-front-national-and-the-future-of-french-foreign-policy

Pietras, Marek, "Poland's Participation in NATO Operations," in Janne Haaland Matláry and Magnus Petersson, eds., *NATO's European Allies: Military Capacity and Political Will*, New York: Palgrave Macmillan, 2013.

Polyakova, Alina, and Spencer B. Boyer, *The Future of Political Warfare: Russia, the West, and the Coming Age of Global Digital Competition*, Washington, D.C.: Brookings Institution, March 2018. As of September 20, 2018:
https://www.brookings.edu/wp-content/uploads/2018/03/the-future-of-political-warfare.pdf

Posen, Barry, *The Sources of Military Doctrine: France, Britain, and Germany Between the World Wars*, Ithaca, N.Y.: Cornell University Press, 1984.

Pressman, Jeremy, *Warring Friends: Alliance Restraint in International Politics*, Ithaca, N.Y.: Cornell University Press, 2008.

Priebe, Miranda, *Fear and Frustration: Rising State Perceptions of Threats and Opportunities*, dissertation, Cambridge, Mass.: Massachusetts Institute of Technology, 2015.

Priest, Dana, "United NATO Front Was Divided Within," *Washington Post*, September 21, 1999. As of September 19, 2018:
https://www.washingtonpost.com/wp-srv/inatl/daily/sept99/airwar21.htm

Putin, Vladimir, "Answers to Journalists' Questions," *Kremlin.ru*, June 6, 2014. As of September 20, 2018:
http://en.kremlin.ru/events/president/news/45869#sel=45:28:r23,46:39:nkw

Rathke, Jeffrey, "Can NATO Deter Russia in View of the Conventional Military Imbalance in the East?" Center for Strategic and International Studies, November 30, 2015. As of September 19, 2018:
https://www.csis.org/analysis/can-nato-deter-russia-view-conventional-military-imbalance-east

Rettman, Andrew, Peter Teffer, Eric Mauice, and Eszter Zalan, "Italy Shields Russia from EU Sanctions Threat," *EUobserver*, October 21, 2016. As of September 20, 2018:
https://euobserver.com/foreign/135593

Reuters, "NATO Must Improve Defences Against a 'More Aggressive' Russia, Says Chief," *The Guardian*, March 18, 2018. As of September 19, 2018:
https://www.theguardian.com/world/2018/mar/18/
nato-must-improve-defences-against-a-more-aggressive-russia-says-chief

Rid, Thomas, and Ben Buchanan, "Attributing Cyber Attacks," *Journal of Strategic Studies*, Vol. 38, No. 1–2, December 2015, pp. 4–37. As of September 19, 2018:
https://www.tandfonline.com/doi/abs/10.1080/01402390.2014.977382

Rivkin, David B., Jr., and Miomir Zuzul, "Trump Is Right on Nord Stream 2," *Wall Street Journal*, July 22, 2018. As of September 20, 2018:
https://www.wsj.com/articles/trump-is-right-on-nord-stream-2-1532289915

"Russia, Turkey Sign Gas Pipeline Deal," *Radio Free Europe/Radio Liberty*, October 10, 2016. As of September 20, 2018:
https://www.rferl.org/a/russia-turkey-putin-erdogan-azerbaijan-aliyev-istanbul/28043861.html

"Russian Embargo: Reactions," *Iceland Monitor*, August 14, 2015. As of September 20, 2018:
https://icelandmonitor.mbl.is/news/politics_and_society/2015/08/14/russian_embargo_reactions

Saideman, Stephen M., and David P. Auerswald, "Comparing Caveats: Understanding the Sources of National Restrictions Upon NATO's Mission in Afghanistan," *International Studies Quarterly*, Vol. 56, No. 1, March 2012, pp. 67–84. As of September 19, 2018:
http://dx.doi.org/10.1111/j.1468-2478.2011.00700.x

Sanger, David E., and Maggie Haberman, "Donald Trump Sets Conditions for Defending NATO Allies Against Attack," *New York Times*, July 20, 2016. As of September 19, 2018:
https://www.nytimes.com/2016/07/21/us/politics/donald-trump-issues.html

Sano, Yoel, "Guest Post: Will Russia Make a Play for Estonia, Latvia and Lithuania?" *Financial Times*, March 23, 2015. As of September 17, 2018:
https://www.ft.com/content/c03befc7-67a8-3e91-93c2-72d865ccb42b

Santora, Marc, "Czech Republic Re-Elects Milos Zeman, Populist Leader and Foe of Migrants," *New York Times*, January 27, 2018. As of September 19, 2018:
https://www.nytimes.com/2018/01/27/world/europe/czech-election-milos-zeman.html

Sarts, Janis, "Russian Interference in European Elections," Hearing Before the Select Committee on Intelligence, U.S. Senate, Washington, D.C., June 28, 2017.

Saunders, Elizabeth N., "War and the Inner Circle: Democratic Elites and the Politics of Using Force," *Security Studies*, Vol. 24, No. 3, September 2015, pp. 466–501. As of September 19, 2018:
http://dx.doi.org/10.1080/09636412.2015.1070618

Schmitt, Eric, "NATO Planes to End Patrol of U.S. Skies," *New York Times*, May 2, 2002. As of September 17, 2018:
https://www.nytimes.com/2002/05/02/us/nato-planes-to-end-patrol-of-us-skies.html

Schöler, Gabriele, *Frayed Partnership: German Public Opinion on Russia*, Gütersloh, Germany, and Warsaw, Poland: Bertelsmann Stiftung and Institute of Public Affairs, Poland, April, 2016. As of September 17, 2018:
https://www.bertelsmann-stiftung.de/fileadmin/files/user_upload/EZ_Frayed_Partnership_2016_ENG.pdf

Schroeder, Robin, and Martin Zapfe, "'War-Like Circumstances'— Germany's Unforeseen Combat Mission in Afghanistan and Its Strategic Narratives," in Beatrice de Graaf, George Dimitriu, and Jens Ringsmose, eds., *Strategic Narratives, Public Opinion, and War: Winning Domestic Support for the Afghan War*, New York: Routledge, 2015.

Scrutton, Alistair, and David Mardiste, "Wary of Divided Loyalties, a Baltic State Reaches Out to Its Russians," *Reuters*, February 24, 2017. As of October 12, 2018:
https://www.reuters.com/article/us-baltics-russia-idUSKBN1630W2

Sharkov, Damien, "Russian Supersonic Jets Force NATO to Scramble Fighter Jets," *Newsweek*, March 25, 2015. As of September 19, 2018:
https://www.newsweek.com/russian-supersonic-jets-force-nato-scramble-fighters-316730

Sheehan, James J., *Where Have All the Soldiers Gone? The Transformation of Modern Europe*, Boston, Mass.: Houghton Mifflin, 2008.

Shirreff, Richard, *War with Russia: An Urgent Warning from Senior Military Command*, New York and London: Quercus, 2016.

Shlapak, David A., and Michael Johnson, *Reinforcing Deterrence on NATO's Eastern Flank: Wargaming the Defense of the Baltics*, Santa Monica, Calif.: RAND Corporation, RR-1253-A, 2016. As of September 19, 2018:
https://www.rand.org/pubs/research_reports/RR1253.html

Simmons, Katie, Bruce Stokes, and Jacob Poushter, *NATO Publics Blame Russia for Ukrainian Crisis, but Reluctant to Provide Military Aid*, Washington, D.C.: Pew Research Center, June 10, 2015. As of September 19, 2018:
https://www.pewresearch.org/wp-content/uploads/sites/2/2015/06/Pew-Research-Center-Russia-Ukraine-Report-FINAL-June-10-2015.pdf

Smoleňová, Ivana, and Barbora Chrzová, eds., *United We Stand, Divided We Fall: The Kremlin's Leverage in the Visegrad Countries*, Prague: Prague Security Studies Institute, November 2017. As of September 19, 2018:
http://ceid.hu/wp-content/uploads/2017/11/Publication_United-We-Stand-Divided-We-Fall-1.pdf

Snyder, Glenn H., "The Security Dilemma in Alliance Politics," *World Politics*, Vol. 36, No. 4, July 1984, pp. 461–495. As of September 19, 2018:
http://www.jstor.org/stable/2010183

Snyder, Jack, and Erica D. Borghard, "The Cost of Empty Threats: A Penny, Not a Pound," *American Political Science Review*, Vol. 105, No. 3, August, 2011, pp. 437–456. As of September 19, 2018:
http://www.jstor.org/stable/41480851

Solovyov, Dmitry, and Andrius Sytas, "Russia Moves Nuclear-Capable Missiles into Kaliningrad," *Reuters*, October 8, 2016. As of September 19, 2018:
https://www.reuters.com/article/us-russia-usa-missiles-confirm-idUSKCN1280IV

"Some 71% of Poles Think Poland Should Defend Baltics in Case of Attack," *Delfi by Lithuania Tribune*, November 10, 2016. As of October 4, 2018:
https://en.delfi.lt/lithuania/defence/some-71-of-poles-think-poland-should-defend-baltics-in-case-of-attack.d?id=72825122

Sperling, James, "Neo-Classical Realism and Alliance Politics," in Mark Webber and Adrian Hyde-Price, eds., *Theorising NATO: New Perspectives on the Atlantic Alliance*, New York: Routledge, 2016, pp. 72–78.

Stacey, Jeffrey A., "A Russian Attack on Montenegro Could Mean the End of NATO," *Foreign Policy*, July 27, 2018. As of September 17, 2018:
https://foreignpolicy.com/2018/07/27/
a-russian-attack-on-montenegro-could-mean-the-end-of-nato-putin-trump-helsinki

Stelzenmueller, Constanze, and Joshua Raisher, *Transatlantic Trends: Key Findings 2013*, Washington, D.C.: German Marshall Fund of the United States, 2013. As of September 18, 2018:
http://www.gmfus.org/publications/transatlantic-trends-2013

Stevenson, Reed, and Aaron Gray-Block, "Dutch Government Falls over Afghan Troop Mission," *Reuters*, February 19, 2010. As of September 19, 2018:
https://www.reuters.com/article/us-dutch-government/
dutch-government-falls-over-afghan-troop-mission-idUSTRE61J0FS20100220

Stokes, Bruce, "NATO's Image Improves on Both Sides of Atlantic: European Faith in American Military Support Largely Unchanged," Pew Research Center, May 23, 2017. As of September 17, 2018:
http://www.pewglobal.org/2017/05/23/natos-image-improves-on-both-sides-of-atlantic

Strüver, Georg, and Tim Wegenast, "The Hard Power of Natural Resources: Oil and the Outbreak of Militarized Interstate Disputes," *Foreign Policy Analysis*, Vol. 14, 2018, pp. 86–106.

Strzelecki, Marek, and Maciej Martewicz, "Gazprom Limits Polish Gas Supplies as Reverse Flows Halt," *Bloomberg*, September 10, 2014. As of September 20, 2018:
https://www.bloomberg.com/news/articles/2014-09-10/
poland-says-gazprom-cut-gas-supplies-via-belarus-ukraine

Stuttaford, Andrew, "Watching the Baltic—'Little Green Men' and Other Concerns," *National Review*, February 8, 2015. As of September 19, 2018:
https://www.nationalreview.com/corner/
watching-baltic-little-green-men-and-other-concerns-andrew-stuttaford

Szabo, Stephen F., *Germany, Russia, and the Rise of Geo-Economics*, London and New York: Bloomsbury Publishing, 2015.

Szczepański, Marcin, *Economic Impact on the EU of Sanctions over Ukraine Conflict*, Brussels, Belgium: European Parliamentary Research Service, PE 569.020, October 2015. As of September 20, 2018:
http://www.europarl.europa.eu/RegData/etudes/BRIE/2015/569020/
EPRS_BRI(2015)569020_EN.pdf

Tasch, Barbara, "The Czech Government Thinks Russian Propaganda Is 'The Biggest Threat Europe Has Been Facing Since the 1930s,'" *Business Insider*, February 3, 2017. As of October 4, 2018:
https://www.businessinsider.com/russian-propaganda-biggest-threat-europe-since-1930-2017-2

Terzuolo, Eric, "Putin Is the Real Winner of the Italian Elections," *The Hill*, March 8, 2018. As of September 19, 2018:
https://thehill.com/opinion/international/377462-putin-is-the-real-winner-of-the-italian-elections

Theiner, Thomas, "Gotland—The Danzig of Our Time," *Euromaiden Press*, March 22, 2015. As of September 17, 2018:
http://euromaidanpress.com/2015/03/22/gotland-the-danzig-of-our-time/

Thompson, Loren B., "Why the Baltic States Are Where Nuclear War Is Most Likely to Begin," *National Interest*, July 20, 2016. As of September 19, 2018: https://nationalinterest.org/blog/the-buzz/why-the-baltic-states-are-where-nuclear-war-most-likely-17044

Thornton, Ron, and Manos Karagiannis, "The Russian Threat to the Baltic States: The Problems of Shaping Local Defense Mechanisms," *Journal of Slavic Military Studies*, Vol. 29, No. 3, 2016, pp. 331–351. As of September 19, 2018: https://www.tandfonline.com/doi/abs/10.1080/13518046.2016.1200359

Tomiuc, Eugen, and Radu Benea, "Russia Objects to Moldovan Call for Removing Troops from Transdniester," *Radio Free Europe/Radio Liberty*, August 23, 2017. As of September 20, 2018: https://www.rferl.org/a/moldova-calls-on-united-nations-russian-troops-transdniester/28693178.html

Tomovic, Dusica, "Russia Threatens Retaliation After Montenegro Joins NATO," *BalkanInsight*, June 6, 2017. As of September 20, 2018: http://www.balkaninsight.com/en/article/montenegro-faces-russian-retaliation-after-joining-nato-06-06-2017

Tomz, Michael, "Domestic Audience Costs in International Relations: An Experimental Approach," *International Organization*, Vol. 61, No. 4, Autumn 2007, pp. 821–840. As of September 19, 2018: https://www.jstor.org/stable/4498169?seq=1#metadata_info_tab_contents

Tsebelis, George, "Decision Making in Political Systems: Veto Players in Presidentialism, Parliamentarism, Multicameralism and Multipartyism," *British Journal of Political Science*, Vol. 25, No. 3, July 1995, pp. 289–325. As of September 19, 2018: https://www.jstor.org/stable/194257?seq=1#metadata_info_tab_contents

Tuschhoff, Christian, "Why NATO Is Still Relevant," *International Politics*, Vol. 40, No. 1, March 2003, pp. 101–120. As of January 8, 2019: https://link.springer.com/article/10.1057/palgrave.ip.8800007

"U.S.-Funded Radio Free Europe/Radio Liberty Fined by Russia," *U.S. News & World Report*, July 5, 2018. As of October 4, 2018: https://www.usnews.com/news/business/articles/2018-07-05/us-funded-radio-free-europe-radio-liberty-fined-by-russia

"U.S.: Russia Uses Energy Supplies 'to Control Ukraine,'" *BBC News*, April 11, 2014. As of September 19, 2018: https://www.bbc.com/news/world-europe-26982173

U.S. Senate Committee on Foreign Relations, *Putin's Asymmetric Assault on Democracy in Russia and Europe: Implications for U.S. National Security—A Minority Staff Report Prepared for the Use of the Committee on Foreign Relations, United States Senate*, Washington, D.C.: U.S. Government Printing Office, S. Prt. 115-21, January 10, 2018a. As of September 19, 2018: https://www.foreign.senate.gov/imo/media/doc/FinalRR.pdf

———, *A Resolution Expressing the Sense of the Senate Regarding the Strategic Importance of NATO to the Collective Security of the Transatlantic Region and Urging Its Member States to Work Together at the Upcoming Summit to Strengthen the Alliance*, Washington, D.C.: U.S. Government Printing Office, S. Res. 557, July 17, 2018b. As of September 20, 2018: https://www.congress.gov/bill/115th-congress/senate-resolution/557/text

Vasquez, John A., *The War Puzzle*, Cambridge: Cambridge University Press, 1993.

Veal, Lowana, "Iceland: Fish Fight Spawned over Ukraine Sanctions," *Al Jazeera*, October 8, 2015. As of September 20, 2018:
https://www.aljazeera.com/indepth/features/2015/10/iceland-fish-fight-spawned-ukraine-sanctions-151007115626775.html

Verseck, Keno, "Is the Czech Republic Moving Closer to China and Russia?" *Deutsche Welle*, January 31, 2018. As of September 19, 2018:
https://www.dw.com/en/is-the-czech-republic-moving-closer-to-china-and-russia/a-42392040

von Loringhoven, Arndt Freytag, "Adapting NATO Intelligence in Support of 'One NATO,'" *NATO Review*, September 8, 2017. As of September 20, 2018:
https://www.nato.int/docu/review/2017/Also-in-2017/adapting-nato-intelligence-in-support-of-one-nato-security-military-terrorism/EN/index.htm

Walker, Peter, "Russian Expansionism May Pose Existential Threat, Says NATO General," *The Guardian*, February 20, 2015. As of September 19, 2018:
https://www.theguardian.com/world/2015/feb/20/russia-existential-threat-british-nato-general

Walt, Stephen M., *The Origins of Alliances*, Ithaca, N.Y.: Cornell University Press, 1987.

Weitsman, Patricia A., "Intimate Enemies: The Politics of Peacetime Alliances," *Security Studies*, Vol. 7, No. 1, Autumn 1997, pp. 156–192. As of September 19, 2018:
http://dx.doi.org/10.1080/09636419708429337

———, "Alliance Cohesion and Coalition Warfare: The Central Powers and Triple Entente," *Security Studies*, Vol. 12, No. 3, Spring 2003, pp. 79–113. As of September 19, 2018:
http://dx.doi.org/10.1080/09636410390443062

Wellman, Phillip Walter, "UK to Nearly Double Troops in Afghanistan After Trump Request," *Stars and Stripes*, July 11, 2018. As of September 19, 2018:
https://www.stripes.com/news/uk-to-nearly-double-troops-in-afghanistan-after-trump-request-1.537044

Williams, Ian, "The Russia–NATO A2AD Environment," *Missile Threat: CSIS Missile Defense Project*, January 3, 2017. As of September 19, 2018:
https://missilethreat.csis.org/russia-nato-a2ad-environment

Winneker, Craig, "For Europe's NATO Allies, Attack on One Isn't Attack on All," *Politico*, June 10, 2015. As of September 17, 2018:
https://www.politico.eu/article/europe-nato-use-of-force-russia-ukraine-pew-research

Wintz, Mark, *Transatlantic Diplomacy and the Use of Military Force in the Post–Cold War Era*, New York: Palgrave Macmillan, 2010.

Xu, Ruike, *Alliance Persistence Within the Anglo-American Special Relationship: The Post–Cold War Era*, Cham, Switzerland: Springer International Publishing, 2017.

Zanotti, Jim, and Clayton Thomas, *Turkey: Background and U.S. Relations*, Washington, D.C.: Congressional Research Service, R41368, August 31, 2018. As of September 20, 2018:
https://fas.org/sgp/crs/mideast/R41368.pdf